Robust SRAM Designs and Analysis

Jawar Singh • Saraju P. Mohanty • Dhiraj K. Pradhan

Robust SRAM Designs and Analysis

Jawar Singh
Indian Institute of Information Technology
Design and Manufacturing
Dumna Airport Road
Jabalpur
India

Saraju P. Mohanty
University of North Texas
Discovery Park, 3940 N. Elm,
 Room F247
Denton
USA

Dhiraj K. Pradhan
University of Bristol
Merchant Venturers Building
Woodland Rd.
Bristol
United Kingdom

ISBN 978-1-4939-0244-6   ISBN 978-1-4614-0818-5 (eBook)
DOI 10.1007/978-1-4614-0818-5
Springer New York Heidelberg Dordrecht London

Jawar Singh dedicated this work to his grandmother Narayni Devi and grandfather late shri Kehari Singh

Saraju P. Mohanty would like to dedicate this work to Uma, parents, and sisters.

Dhiraj K. Pradhan would like to dedicate this work to his family

Preface

During the past decades, Complementary Metal Oxide Semiconductor (CMOS) technology has emerged as the dominant fabrication method and became the exclusive choice for semiconductor memories. Particularly, Static Random Access Memories (SRAMs) that play a significant role in the memory hierarchy of a modern computer system and continue to be a critical component across wide range of microelectronics applications from consumer wireless to high performance server processors, multimedia and System-on-Chip (SoC) products. SRAM bitcells in general are made of minimum geometry devices for high density to keep pace with CMOS technology scaling; as a result, they are the first to suffer from technology scaling induced side effects. Therefore, several alternate SRAM bitcell circuits and architectures have recently been proposed to meet the nano-regime challenges such as low-power, process variation and soft errors. Emphasis is also given on alternate devices such as Tunnel Field Effect Transistors (TFETs) based SRAM bitcells for low power applications.

The objective of this book is to provide a systematic and comprehensive insight which aids the understanding of SRAM bitcell circuits, architectures, and design and analysis techniques. The nano-regime challenges such as low-power, process variation and soft errors are the core issues considered while designing and analyzing the SRAM bitcells in depth. Robust SRAM designs and analysis techniques show circuit and embedded system designers, researchers, and engineers various aspect of design and analysis of SRAM bitcell circuits and arrays. The above concepts are further elaborated to provide in depth guidance to large cache design needed in embedded and portable systems. The text provides alternative topologies to six-transistor (6T) SRAM which are more robust when implemented using state-of-the-art nano-scale CMOS technology. Topologies for low-power SRAM bitcells are classified on the basis of their robustness and elaborated along with their merits and de-merits. Various quality metrics are discussed to meet the small and large sized cache memories.

The content of this book is directed to nano-scale VLSI design engineers, graduate students in electrical engineering, and computer scientists who are about to start their research in SRAM design. It is an important source for engineers who intended to develop and understand the different aspects of SRAM. The text assumed that readers have basic knowledge and familiarity with electronic circuits and devices. The goal of this book is to train the VLSI design engineers and students to design SRAM and cache architecture rationally.

This book provides a sufficient amount of fundamentals to become familiar with the terminology of the SRAM design and analysis. The main objective is to achieve in depth knowledge in few topics such as operation, design and analysis of CMOS and TFET based SRAMs. The book is organized in six chapters: (1) Introduction to SRAM, (2) Design Metrics of SRAM Bitcell, (3) Single-ended SRAM Bitcell Design, (4) 2-Port SRAM Bitcell Design, (5) SRAM Bitcell Design using Unidirectional Devices, and (6) NBTI and its effect on SRAM. The introductory description of the SRAM serves as a basis for understanding the importance of SRAM in memory hierarchy and need of basic building blocks for the realization of cache memories. Basic operations and static and dynamic stability analysis for small and large sized caches show how conventional methods do not provide adequate data points for stability analysis. Case study of single-ended six-transistor and 2-port SRAM bitcells based cache modules show the complete design flow of SRAMs. The implications in realizing of SRAM using unidirectional devices such as TFETs are studied in details and two different SRAM bitcells using TFETs are compared with standard six-transistor CMOS SRAM bitcell. Finally, the impact of Negative Bias Temperature Instability is studied on different SRAM configurations.

Authors of this book are grateful to all the people without whom the work could not have been accomplished. Saraju P. Mohanty will like to acknowledge Dean College of Engineering at UNT. He will acknowledge CSE department chairman, colleagues, and staff for their support. He will like to acknowledge all his past and current students.

Authors would like to thank invaluable support from our families during the preparation of this book. Jawar Singh would like to express his appreciation to his wife Jyoti, their daughter Jeevika and son Siyon for their understanding and support. Saraju P. Mohanty would like to express thanks to Uma, parents, and sisters.

Jawar Singh
Saraju P. Mohanty
Dhiraj K. Pradhan

Contents

Chapter 1
Introduction to SRAM

1.1 CMOS Technology Scaling

CMOS technology scaling driven by Moore's law has rapidly increased VLSI designs performance by five orders of magnitude in last four decades. According to Moore's law, which was historically formulated in 1965, states the doubling of the number of transistors per generation on an integrated circuit almost every 2 years (usually 18–24 months) [80]. Since that time, Moore's law has become the fundamental guideline for the semiconductor industry to scale down the process technologies of the future generations. The semiconductor industry is understandably desperate to see the pace of Moore's law continue, and that pace is dependent on the technology that can create those ever-shrinking transistors and to overcome the associated challenges of technology scaling. He also stated that the manufacturing cost per function in microprocessor would drop-off exponentially for future generation technologies.

In general, scaling the minimum feature size, length and width by about 30% (Moore's magic number) for each new technology generation, theoretically yields the following:

1. Doubles the device density, while area lowers by $(0.7*Y \times 0.7*X) \sim 50\%$, packing in more devices in the same area, which effectively lowers the cost per transistor;
2. Reduces the total capacitance by 30% which allow gate delays to decrease by 30%, resulting in increase in operating speed up to 43%;
3. Accordingly the power consumption $(Power \propto CV^2 f)$ should decrease for a given circuit by 30–65% due to smaller transistors and lower supply voltage [16].

Figure 1.1, illustrates the CMOS technology scaling. This 30% magic number dictates the next generation of CMOS technology according to Moore's low. The idea of technology scaling is very attractive. The Semiconductor industry has worked very aggressively to continue this trend of technology scaling, however, the pace of this aggressive scaling has been slow down in the recent past. In order

Fig. 1.1 Illustration of
CMOS technology scaling
for future generations

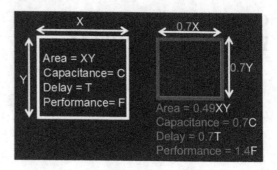

to drive next generation technology node from the Moore's magic number, if the current technology node is 65 nm then the next technology node is (65*0.7) 45 nm. Similarly, all other technology generations have been derived.

Scaling supply voltage drastically reduces the dynamic power due to quadratic relation with supply voltage and static power. However, simply lowering V_{DD} will increase delay, so the device threshold voltage, V_{TH}, must also decrease in order to maintain the drive current. Lowering V_{TH} leads to an exponential increase in leakage power. Moreover, minimum feature sized and closely matched devices matter significantly, particularly when designing Static Random Access Memories (SRAMs), therefore, they are the first to suffer from the exponential trends of scaling. The continued scaling of CMOS technology has resulted several problems these include process induced variations, soft errors, transistor degradation due to ageing etc. However, these problems were less severe in the earlier generations.

1.2   Why SRAM?

The origination of the concept of the MOSFET based memory was first commercialized and perfected in the seventies. Robert Dennard of IBM envisaged the dynamic memory cell using a single MOSFET and a capacitor in 1968 [30]. The first MOSFET based dynamic random access memory (DRAM) chip with 2k-bits was developed in 1971 with several process improvements in leakage control. However, DRAM performance has not kept the pace with the performance of the processors from the very beginning [29, 42] due to long access time and more power hungry. The dynamic nature of DRAM requires that the memory must be refreshed periodically so as not to lose the content of the memory cells.

The growing gap between the processors and the DRAM performance has dictated the need of different levels of memory hierarchy in the processor architectures. The memory hierarchy ranges from high-performance, small sized but expensive on-chip memories to slower, large sized but inexpensive off-chip memories such as DRAM, magnetic or optical memories. To meet the system performance requirements, the processor tries to keep frequently used data and

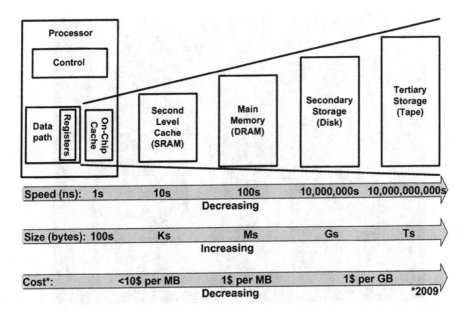

Fig. 1.2 Typical memory hierarchy of a modern computer system

instructions closer to itself, that is, in the faster on-chip memory, which is referred as "cache" memory. A typical memory hierarchy of a modern computer system is depicted in Fig. 1.2. The on-chip cache memories are often called L1, L2 and even L3. The different levels of cache memories are static random access memories (SRAMs) and they dominate the memory hierarchy in performance but they are often integrated in a lesser capacity due to area limitations and the high cost per bit. The speed and the cost per bit decrease as one moves from registers to tertiary storage, however, data storage capacity increases.

SRAMs continue to be critical component across a wide range of microelectronics applications from consumer wireless to high performance server processors, multimedia and System on Chip (SoC) applications. Modern high performance processors and SoC application demands more on-chip memory to meet the performance and throughput requirements. However, it is also not feasible to embedded large amount of memory needed into the chip due to area limitations and the high cost per bit. Figure 1.3 shows the increasing trend of on-die cache memory for different processors based on different technology nodes. It is also projected that the percentage of embedded SRAM in SoC products will increase further from the current 84% to as high as 94% by the year 2014 [48]. Furthermore, their is a huge demand of cache memory in modern computer systems as microprocessors design paradigm has been shifted to multi-core architectures. As shown in Fig. 1.3, the amount of on-die cache in Montecito, Dual Core, Intel processor has increased significantly as compared to Xeon single core processor.

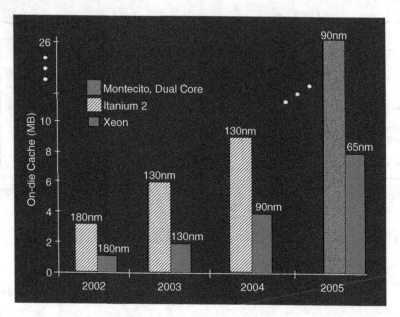

Fig. 1.3 The amount of on-die cache memory for different processors based on different technologies

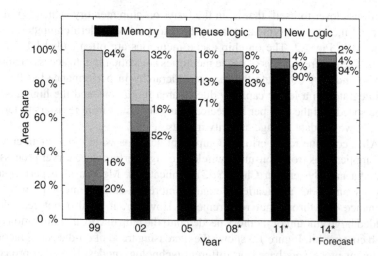

Fig. 1.4 Typical trend of memory and logic area on an system-on-chip (SoC) die [48]

Typical trend of embedded memory and logic area on a system-on-chip (SoC) is also shown in Fig. 1.4. This trend shows that how the share of SRAM on a die has drastically increased from 20% in 1999 to 94% as forecast in 2014. This growing trend is mainly to provide faster access by eliminating the delay across the chip interface. Also embedded memories are designed with rules more aggressive than the rest of the logic on a SoC die, therefore, they have dense packing which makes them more prone to manufacturing defects. This trend has mainly grown due to ever increased demand of performance and higher memory bandwidth requirement to minimize the latency, therefore, larger L1, L2 and even L3 caches are being integrated on-die. Hence, it may not be an exaggeration to say that the SRAM is a good technology representative and a powerful workhorse for the realization of modern SoC applications and high performance processors. In addition, SRAM scaling signifies the huge potential of decreasing the cost per function in microprocessors as well.

1.3   SRAM Architecture

An SRAM cache consists of an array of bi-stable memory bitcells along with peripheral circuitries, such as address (row and column) decoders, sense amplifiers, write drivers and bitline pre-charge circuits etc. Peripheral circuitries enable reading from and writing into the array. A classic SRAM memory architecture is shown in Fig. 1.5. The memory array consists of 2^n words of 2^m bits each. An SRAM array is composed of millions of identical bitcells. For example, a 32 Mb cache memory is composed of 33,554,432 bitcells, a number so great that even an exceptionally rare event can have a noticeable impact on product yield. As a result, small improvement in reliability, performance and saving in static power will have a great impact on the entire processor or SoC product. Therefore, optimization of the SRAM bitcell designs for a target application is an active area of research. In high performance processors, operating speed and bitcell area are the prime concern in order to have high density caches, while, maintaining an adequate reliability. However, in energy constrained applications such as sensor nodes or medical implants, energy efficiency and reliability are the main issues.

A memory bitcell is a circuit capable of storing a single bit of information – "1" or "0". They share a common wordline (WL) in each row and a bitline pairs (BL, complement of BL) in each column of an SRAM array. The dimensions of each SRAM array are limited by its electrical characteristics such as capacitances and resistances of the bitlines and wordlines used to access bitcells at uniform delay in the array. Memory arrays are organized such that the horizontal and vertical dimensions are of the same order of magnitude. Therefore, large size memories may be folded into multiple blocks with limited number of rows and columns. After folding, in order to meet the bit and word line capacitance requirement each row of the memory contains 2^k words, so the array is physically organized as 2^{n-k} rows and 2^{m+k} columns. Every bitcell can be randomly addressed by selecting the appropriate wordline (WL) and bitline pairs (BL, complement of BL), respectively, activated by the row and the column decoders.

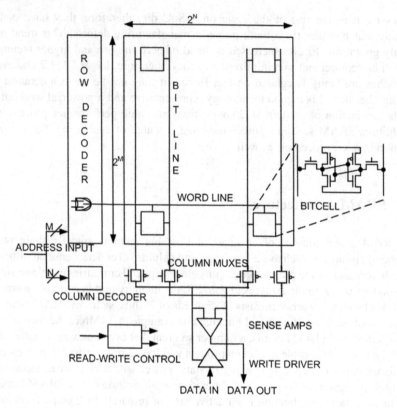

Fig. 1.5 A general SRAM array structure

1.3.1 SRAM Bitcell

An SRAM bitcell is the basic building block of the SRAM array, as shown in the inset of Fig. 1.5. Each bitcell circuit is capable of storing single bit of information. It provides non-destructive read operation, write capability and data storage as long as the SRAM bitcell is powered up. A standard six transistor (6T) SRAM bitcell consists of two cross coupled inverters and two access transistors connected to each data storage node. The inverter pair forms a latch and holds the binary information. True and complimentary version of the binary data are stored in the storage nodes. The access transistors allow access to data storage nodes during read and write operations and also provides isolation from the other neighbouring circuits during hold state. The bitcells are accessed horizontally by asserting the wordline during read and write operation. When wordline of a row is asserted 'HIGH', all the memory bitcells in the selected row become "active" and can be ready for read and write operations. To decode m wordlines, one needs $log_2 m$ address bits. An SRAM bitcell has three modes of operation: read, write and standby; or in other words, it can be in three different states such as reading, writing or data retention.

1.3.2   Address Decoders

To implementing an N-word memory where each word is M bits wide, a general approach is to arrange the memory words in a linear fashion. In order to read or write, each word is selected with N select lines to access N independent locations. However, this approach seems very simple and works well for small memories, but puts in trouble if N is large (for larger memories). For instance, in a 32 Mb (2^{25}) word-oriented SRAM with a 32-bit (2^5) word width, N = 2^{20} (N = 1,048,576) select lines are needed – one for every word. However, for a 32 Mb bit-oriented SRAM, N becomes 2^{25} (33,554,432). Hence, a large number ($\sim$1 million) of select lines or signals are needed to address this word-oriented memory, if arranged in a linear fashion. As a result, this (linear) approach leads to insurmountable wiring (interconnects) and packaging requirement. In order to reduce the number of select lines or in other word the number of interconnects, a address decoder is inserted. Address decoder allows the number of select lines in the SRAM to be reduced by a factor of log_2N, where N is the number of independent locations. For instance, in a 32 Mb (2^{25}) word-oriented SRAM with a 32-bit (2^5) word width, this approach reduces the number of select lines from $\sim$1 million to 20 ($log_2 2^{(25-5)} = 20$) address bits A0, A1,...A19. This SRAM can be orginzed in 32 blocks each of which has 1,024 rows and 1,024 columns

There are two types of decoders used in the SRAM, that is, row decoder and the column decoder. The design of these decoders has substantial impact on the SRAM performance and power consumption. Row decoders are needed to select one row of wordlines out of a set of rows in the array according to address bits. While column decoder select the particular bitline pairs out of the sets of bitline pairs in the selected row. A fast decoder can be implemented using AND/NAND and OR/NOR gates. These decoders can be implemented in two different styles, namely static and dynamic. The choice of a design styles depends on the SRAM area, performance, power consumption and architectural considerations. The static NAND-type structure can be chosen because of its low power consumption during the decoded row transitions. While dynamic structure can be chosen because of its speed and power improvement over the static NAND gate based decoder.

For large SRAM arrays where total address space is A0, A1,...A19 address bits. In this address space, row decoder requires 10 bits row decoder and 10 bits for column decoder. For the implementation of a row decoder 10-input NOR gate is needed per row. This poses different challenges such as large fan-in which has negative impact on the performance, power dissipation etc. Therefore, splitting a large gate into small logic lavels in general produces a faster, area efficient and cheaper implementation. However, for small single-block memories single stage row decoders are good choice. Today most memories split the row decoder into several blocks decoded by separte decoder stages. The split or multi stage decoder approach has proven to be more efficient for larger memories, it reduces the number of transistor, fan-in, power and loading on the address input buffers. The multi stage decoder structures are classified into two broad categories such as Divided Wordline (DWL) [43] and Hierarchical Word Decoding (DWD) [119] structures.

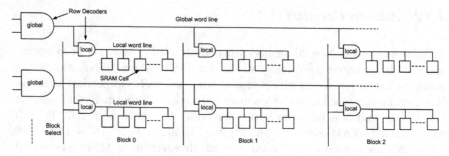

Fig. 1.6 Divided wordline row decoder [43]

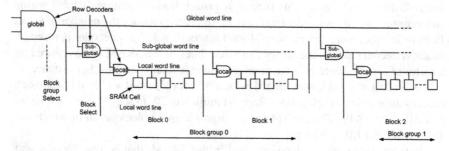

Fig. 1.7 Hierarchical Word Decoding (DWD) scheme [119]

Figure 1.6 shows the DWL structure in which SRAM is partitioned into blocks. In order to read or write a block, local wordline is activated when both global wordline and block select are asserted. Since, only one block is activated at a time for read or write operation, as a result the DWL structure reduces both wordline delay and power consumption. For high density and large SRAMs greater than 4 Mb, the hierarchical wordline decoding structure, as shown in Fig. 1.7 was proposed to cope with increased delay and power consumption.

1.3.3   Precharge Circuit

In all SRAMs, for each column in the bitcell array there is a bitline pair (BL and complement of BL). Each pair of bitlines is connected to a precharge circuit. The function of this circuit is to pull-up the bit lines of a selected coulmn to V_{DD} level and perfectly equalized them before the read or write operation. A typical precharge circuit is shown in Fig. 1.8a. It is composed of a pair of PMOS transistors and a precharge circuit enable signal ($\overline{PC}$), when both the transistors are in ON state, that is, $\overline{PC}$ is active low, bitlines (BL and complement of BL) are connected to V_{DD}. Recently, two transistor precharge circuit shown in Fig. 1.8a has been replaced by a three transistor configuration as shown in Fig. 1.8b. In this precharge circuit

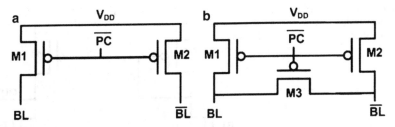

Fig. 1.8 Precharge circuits for SRAM array

transistor M1 and M2 connect the bitlines (BL and complement of BL) to V_{DD} for pull-up, while transistor M3 equalizes both the bitlines. In precharge circuit PMOS transistors are commonly used because they have good V_{DD} passing capacity.

1.3.4   Sense Amplifiers

Sense Amplifiers (SA) are one of the most important peripheral circuits in the CMOS Static Random Access Memories, and become a separate class of circuits in the literature. The primary function of a SA in SRAMs is to amplify a small differential voltage developed on the bitlines during read access and translate it to full swing digital output signal. A small differential voltage is developed by pulling down one of the precharged bitline by the read access bitcell. Due to small bitcell size and large bitlines capacitance, time required for read operation increases significantly, or in other words, read access time increases. These circuits have strong impact on the read access time of a memory (or performance), as they are used to retrieve the stored data in the memory array by amplifying small signal variations on the bitlines.

The design of fast, low-power and robust SA circuits is a challenge, due to the fact that in modern memory design bitlines exhibit a significantly large capacitance. A large number of bitcells per bitlines are generally embedded in modern SRAMs to increase the array density, increased sensitivity to process variations, environmental conditions and device mismatch. Hence, these challenges set limits in the sensing speed, robustness and introduces extra signal delay. The sense amplifier design, furthermore, depends on the timing requirements and layout constraints of the memory system. To alleviate some of the above challenges, sense amplifiers are often employ devices with non-minimum length and width. A sense amplifier is characterized by the following parameters: gain A, sensitivity S, current and voltage offsets V_{off} and I_{off}, common mode rejection ratio $CMMR$, rise time t_{rise}, fall time t_{fall}, and sense delay t_{sense}.

Figure 1.9 shows a commonly used current-mirror differential sense amplifier. The differential sensing is widely used to reject the common-mode noise that may present on both the bitlines. This noise may be induced on both SA inputs or bitline pair (BL and complement of BL) due to power spikes, capacitive coupling between

Fig. 1.9 Commonly used
current-mirror differential
sense amplifier

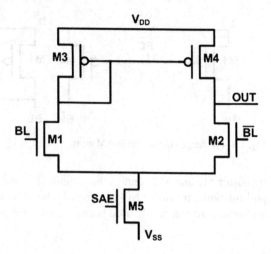

Fig. 1.9 Commonly used current-mirror differential sense amplifier

bitlines and between wordline and bitlines. This noise then attenuated by the value of CMMR and the true differential signal is amplified. Sensing operation in this commonly used current mirror differential sense amplifier begins with setting up the SA operating point by precharging and equalizing of the both the SA inputs, that is, the bitline pair (BL and complement of BL) of a selected column. Once both bitlines are precharged and equalized, wordline WL is asserted to activate the read-accessed bitcells that started build-up of the differential voltage on the bitlines. Once the differential voltage exceeds the sensitivity of the SA or overcomes the offset of SA, Sense Amplifier Enable (SAE) signal is issued to trigger the SA which amplifies the differential voltage caused by the bitcell (connecting one of the bitline to ground via one of the access and pull-down transistor) on the bitlines to full-swing digital output level. The read operation completes with the de-assertion of the SAE and WL. The gain A of a current-mirror sense amplifier is given by Eq. 1.1, and it is typically set to around ten.

$$A = -g_{mM1}\left(r_{o2}\|r_{o4}\right) \tag{1.1}$$

Where g_{mM1} is the transconductance of transistor M1, and r_{o2} and r_{o4} are small-signal output resistance of transistor M2 and M4, respectively. The gain A is directly related to the width of transistor M2 and M4 and can be increased by widening these transistor or by increasing the biasing current.

Another, most commonly used sense amplifier is latch-type, as shown in Fig. 1.10. This amplifier is comprised of two cross coupled inverters and a transistor M5 which isolates it from the bitlines and prevents the discharge of bitline on the '0' storage node. Sense operation in this type of sense amplifier begins with biasing it in the high gain metastable state by precharging and equalizing its inputs (or both bitlines). Additional pass transistors are commonly used to isolate the bitlines connected to precharge circuits. Once a differential voltage is developed

Fig. 1.10 Latch-type sense amplifier

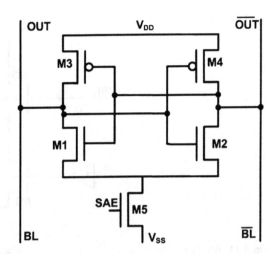

Fig. 1.11 Transmission gates based write driver circuit

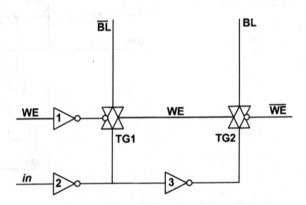

on the bitlines exceed the sensitivity of the sense amplifier, Sense Amplifier Enable (SAE) signal is enabled and the bitlines isolation pass through transistors are turned off. The feed-back mechanism of this amplifier immediately picks up the differential voltage and drives the outputs to the full swing differential voltage.

1.3.5 Write Drivers

Write drivers are used by a group of columns in an SRAM array to control the bitline pair (BL and complement of BL) during write operation. As the bitline pair is precharged to V_{DD} before every operation, the write driver has just to act the pull down one of the two bitlines below the write margin of the SRAM cell during write operation according to the input data. There are different types of write drivers commonly used in SRAM array, some of the typical write driver circuits are discussed below. The write driver circuit shown in Fig. 1.11 comprises

Fig. 1.12 Pass gates based
write driver circuit

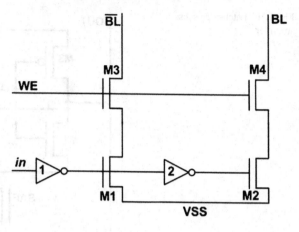

Fig. 1.13 AND gates based
write driver circuit

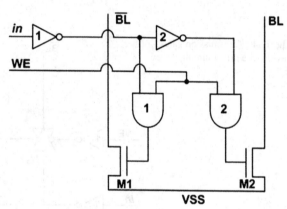

of two transmission gates (TG1 and TG2) and three inverter buffers (INV-1,INV-2
and INV-3). Inverter buffer 2 and 3 writes the data *in* to the bitlines (BL and
complement of BL) via TG1 and TG2. Write driver is enabled by the Write Enable
(WE) signal and drives the bitlines to data *in*. The TG1 and TG2 are activated by WE
and its complementary $\overline{WE}$, as a result one of the bitline is discharged through the
NMOS transistor of either inverter (INV-2 or INV-3). A successful write operation is
ensured in the SRAM bitcell by pulling down the one of the bitlines well below the
write margin of the bitcell. Therefore, the pull down strength of INV-2 and INV-3
plays a significant role in the write operation.

The write driver shown in Fig. 1.12 writes data through two stacked NMOS
transistors, that is, M1, M3 and M2, M4, which form two pass-transistor AND gates.
Write driver is enabled by the WE signal which activates the transistor M3 and M4,
while data *in* enables the transistor M1 or M2 depending upon the input data through
inverter buffers (INV-1 and INV2). When WE is enabled, one of the bitline (BL and
complement of BL) is discharged from precharged level to ground through one of
the transistor M1 or M2 depending upon the input data. Figure 1.13 shows another

implementation of a write driver. In this write driver, when WE is asserted, the combination of WE and data *in* turn on one of the pass transistor M1 or M2. As a result activated transistor (either M1 or M2) discharges the one of the bitline (BL and complement of BL) to the ground level. In SRAM architecture, only one write driver is needed per coulumn in an SRAM array, as a result, write drivers can be upsized to discharge large capacitive bitlines for a successful and faster write operation.

1.4 SRAM Design Issues and Challenges

Unfortunately, the pace of ever-shrinking transistors has brought up many difficult challenges that threaten to stop the exponential trend of doubling the number of transistors per generation and drop-off of manufacturing cost. Perhaps some of the greatest challenges of technology scaling are exponential increase of leakage power, power density and device mismatch (process variation). As technology scales down leakage current increases exponentially and reliability goes down significantly due to poor stability noise margins and process variation. These technology scaling-induced side effects are further exacerbated by reduced supply voltage introduced in order to achieve energy efficiency or low-voltage operation. Achieving low-voltage operation in SRAM faces several challenges such as originating from process variation, related to bitcell read and write stability, sensing, and inefficient Computer Aided Design (CAD) methodologies.

Figure 1.14 shows the comparison of normalized read Static Noise Margin (SNM) and leakage current of a 6T SRAM bitcell for different technology nodes. The minimum feature sized devices with cell ratio ($\beta = 2$), is used for simulation using Predictive Technology Models (PTM) [88, 123]. It can be seen from the Fig. 1.14 that the read SNM of 6T bitcell is gradually decreasing with technology

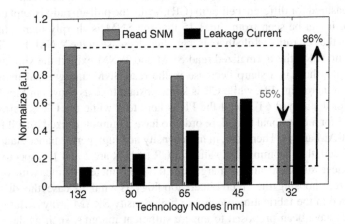

Fig. 1.14 Comparison of read SNM and leakage current of standard 6T SRAM bitcell for different technology nodes

scaling, while the leakage current is exponentially increasing. Moving from 130 to 32 nm technology node, there is 55% reduction in the read SNM while there is 86% increase in leakage current. Therefore, alternative array structure and bitcell design topologies or methodologies are needed for nano-regime technologies that provide low standby power and higher stability margins. In the line of array structures, interleaved strcuture is commonly used while non-interleaved strcuture has been atteracted attention recently. Recently, several SRAM bitcell topologies have been proposed those have very good read SNM and write noise margin and capability to operate at low voltages.

1.4.1   Conflicting Device Size Requirements

The standard 6T SRAM bitcell design space is continuously narrowing down due to shrinkage in device dimensions, attempting to achieve the high density and high performance objectives of on-chip caches. The SRAM bitcell stability and write-ability margins are further degraded by supply voltage scaling. The degradation in margins is mainly due to conflicting read and write requirements of the device size in the 6T bitcell. As both read and write operations are performed via the same pass-gate (NMOS) devices, that is, M_1 and M_2, as shown in Fig. 1.21. For a better read stability (read SNM), both pull down devices, M_4 and M_6 of the storage inverters must be stronger than the pass-gate devices, M_1 and M_2. While for write operation the opposite is desirable, that is, pass-gate devices, M_1 and M_2, must be stronger than pull up devices, M_3 and M_5, to achieve the better write-ability, that is weak storage inverters and strong pass-gate devices. This conflicting trend is also observed when we simulated the read SNM and write noise margin for different cell ratios ($\frac{W_4/L_4}{W_1/L_1} = \frac{W_6/L_6}{W_2/L_2}$) and pull up ratios ($\frac{W_3/L_3}{W_1/L_1} = \frac{W_5/L_5}{W_2/L_2}$).

Figure 1.15 shows the standard 6T SRAM bitcells' normalized read SNM and WNM measured for different cell ratio (CR), while the pull-up ratio is kept constant (PR = 1). It can be seen from Fig. 1.15 that the SNM is sharply increasing with increase in the cell ratio, while there is a gradual decrease in the WNM. For different pull-up ratio (PR), the normalized read SNM and WNM exhibit the similar trend. For example, there is a sharp decrease in the read SNM and gradual increase in WNM with increasing PR, while CR is kept constant to 2, as shown in Fig. 1.16. In general, for a standard 6T bitcell the PR is kept to 1 while the CR is varied from 1.25 to 2.5 for a functional bitcell, in order to have a minimum sized bitcell for high density SRAM arrays. Therefore, in high density and high performance standard 6T SRAM bitcell, the recommended value for CR and PR are 2 and 1, respectively.

To achieve sufficient bitcell margins in the nano-meter regime, with optimum device size, choosing the right bitcell topology is vital to ease the difficulties encountered in the fabrication process in high density SRAM arrays. Many design techniques have been proposed to ensure sufficient margins, such as de-coupling of read and write operations by modifying the bitcell topology. For example, in

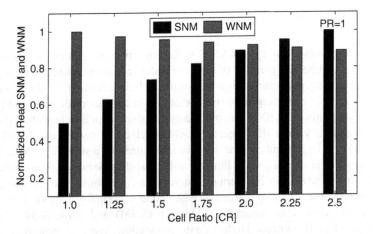

Fig. 1.15 Normalized read SNM and Write Noise Margin (WNM) of a standard 6T SRAM bitcell for different cell ratio (CR), while the pull-up ratio (PR) is fixed to 1

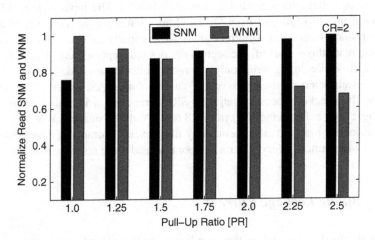

Fig. 1.16 Normalized read SNM and Write Noise Margin (WNM) of a standard 6T SRAM bitcell for different pull up ratio (PR), while the cell ratio (CR) is fixed to 2

an 8T SRAM bitcell topology [23–25, 103], as shown in Fig. 1.25, two transistors were added to create an isolated read-port or read-buffer. The isolated read-port mechanism offers a static-noise-margin-free read operation. In these topologies read and write operations were performed via separate pass-gate devices, thus two are entirely isolated. It avoids the conflicting requirement of sizing the pass-gate devices, which exist in the standard 6T bitcell to achieve a delicate balance between read stability and write-ability. Hence, it widens the bitcell optimization space to a large extent by ensuring sufficient bitcell margins.

1.4.2   Process Variation

However, aggressive scaling of CMOS technology presents a number of distinct challenges for embedded memory fabrics. For instance, smaller feature sizes imply a greater impact of process and design variability, including random threshold voltage (V_{TH}) variations, originating from the fluctuation in number of dopants and poly-gate edge roughness [75, 104]. The process and design variability leads to a greater loss of parametric yield with respect to SRAM bitcell noise margins and bitcell read currents when a large number of devices are integrated into a single die. Predictions in [10] suggest the variability will limit the voltage scaling because of degradation in the SNM and write margin. Furthermore, increase in device mismatch that accompanies geometrical scaling may cause data destruction at normal V_{DD} [19]. Therefore, a sufficiently large read Static Noise Margin (SNM) and Write-Noise Margin (WNM) in a bitcell are needed to handle the tremendous loss of parametric yield.

The stability of a 6T SRAM bitcell under process variation can be verified by examining its butterfly curve obtained by voltage transfer characteristics (VTCs) and inverses voltage transfer characteristics (VTCs^{-1}). The input–output voltage relations from Q to QB and from QB to Q are plotted on the same set of axes, assuming that bitlines and wordlines are biased properly, as shown in Fig. 1.17a, b. During read, three roots of intersection are desired, representing bistability of a SRAM cell. While during write, only one root of intersection is desired in order to flip the bitcell deterministically to one of the two data states, as set by the bitline polarity. The effect of process variation on SRAM read and write stability is evident from Fig. 1.17a, b. Reducing V_{DD} from 0.8 to 0.4 V in the simulation of an SRAM bitcell using 32-nm predictive technology files reveals a dramatic degradation in read and write butterfly curves as result poor read and write noise margins.

1.4.3   Bitline Leakage Current

During the read access when the wordline is activated and access devices are enabled, the bitcell read-current, I_{read}, is the current sunk from the precharged bitline (BLB) (Fig. 1.18) connected to the bitcell node (Q) holding '0' via access device (M_2), as shown in Fig. 1.21. At lower operating voltages, I_{read} is significantly reduced due to lower gate-drive voltage, which implies that the read access time increases substantially. This is undesirable from a performance point of view, but even more importantly it affects the ability to correctly sense the data.

The reduced read-current (I_{read}) and increased aggregate leakage current from the unaccessed bitcells connected to the same bitlines can make conventional data sensing impractical. The reduced I_{read} and increased aggregate leakage current also restrict the number of bitcells per bitline and makes the SRAM array less effecient. Generally, in standard 6T SRAM bitcell differential sense-amplifier is used for detecting a small droop on one of the bitlines, BL or BLB, differentially with respect to the other bitline during the read cycle. A small droop in the BL or BLB is

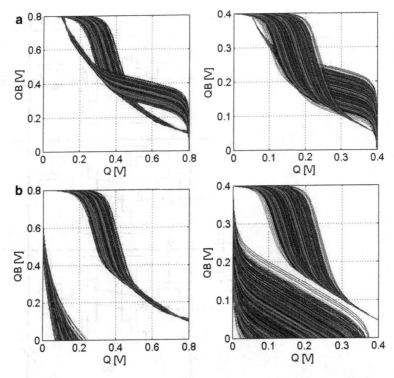

Fig. 1.17 Read and write stability examination under process variation. (**a**) Statistical butterfly curves simulated with intra-die variation at the nominal process corner in 32 nm technology node for read operation at $V_{DD} = 0.8$ and $V_{DD} = 0.4$ V. (**b**) Statistical butterfly curves simulated with intra-die variation at the nominal process corner in 32 nm technology node for write operation operation at $V_{DD} = 0.8$ and $V_{DD} = 0.4$ V

mainly due to the I_{read}, while other bitline is expected to dynamically remain high. However, the aggregate leakage currents on this other bitline depend on the fact that the data stored in the unaccessed bitcell may also be sunk to this bitline, and this makes the differential sensing difficult. Furthermore, if the aggregate leakage currents exceed the I_{read} than there may be an erroneous read operation. Figure 1.18 shows the worst-case bitline leakage current scenario where the data in all of the unaccessed bitcells is such that the aggregate leakage current should nominally be high, so that the droop in voltage in both the bitlines is indistinguishable.

1.4.4 Partial Write Disturbance

The read Static Noise Margin (SNM) problem due to a raise in potential of the node voltage holding '0' is well understood in standard 6T SRAM bitcells. However, the stability problems also arise during a write operation to an unselected column,

Fig. 1.18 Schematic diagram
of a column with a standard
6T SRAM bitcell showing the
I_{read} and aggregate leakage
currents in the worst-case
scenario

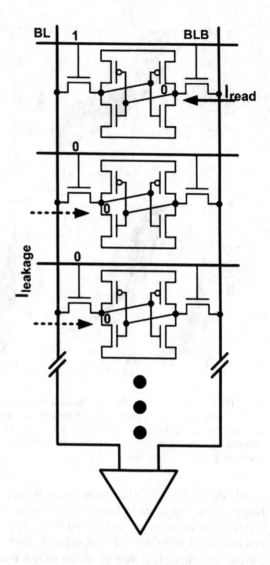

when the wordline is activated while both bitlines are precharged to V_{DD}, which
is a situation that produces equivalent bias conditions to a read operation. The
read stability problem is eliminated by adding a read-port or providing separate
read and write ports, such as, an 8T SRAM bitcell at the cost of increased area
overhead. While the only true method to eliminate bitcell disturbance during such
a partial write operation or partial write disturbance (PWD) is that it requires the
column select functionality within the array to be disallowed. The PWD problem is
explained in detail in Chap. 2.

In dual-port (1-read/1-write) SRAM bitcells, the simultaneous read and write
operations and PWD further exacerbate the stability problems. Therefore, it requires modifications in the SRAM array organization. This modification is just as

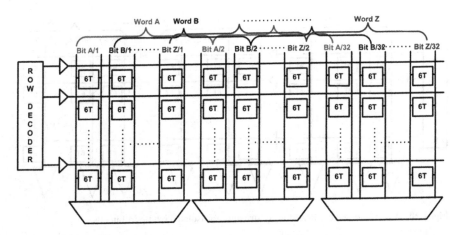

Fig. 1.19 An interleaved SRAM array structure

important as the bitcell itself. Such array-level changes are necessary to achieve the full stability and scaling benefits of an 8T SRAM bitcell, such as, SNM free operation and improved WAM, and it is shown how they are tightly related to the device feature sizes.

1.4.5 Soft Errors

Soft errors (or single-event-upsets) can also corrupt the data in SRAM cells due to radiation. When neutron from space or alpha particles from packaging materials penetrates a silicon wafer, they generate charge particles that perturbs the active (data storage) nodes of a memory cell causing to flip. Thus, failure rate increases with reduction in supply voltage due to reduction in stored charged on the internal data storage nodes. These soft errors can be addressed by providing the error correcting codes which requires redundant SRAM cells in each word. The error correcting codes in SRAM add latency to both write access (for encoding) and read access (for detection and correction). Redundancy goes significantly high if the number of errors corrected and detected is more than 1 bit. Therefore, multi-bit errors caused by soft error phenomena can be avoided by interleaving multiple words onto same physical row [100].

Figure 1.19 shows a conventional interleaved SRAM array structure, where each bit of Z words is laid-out contiguously in a row. To read/write a particular accessed word, the corresponding word-line is activated by the row decoder. Since all bitcells in a row share the same word-line, all the words bitcells are activated simultaneously by the word-line drivers. Accordingly, all the bit-lines for an accessed word are connected to the sense-amplifiers or write-drivers by the column multiplexors. For instance, to read an accessed word B from the first row, remaining words are also activated by the first word-line, hence, these words are referred as

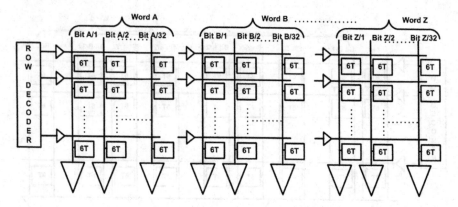

Fig. 1.20 A non-interleaved SRAM array structure for isolated read-port bitcells

half-selected words. These half-selected words are susceptible to read-destructive failure and poor read SNM conditions. The similar problems also exist during write operation in half-selected words, referred as Partial Write Disturbance (PWD). However, the use of read-buffer (isolated read-port) bitcell topologies eliminates the poor read SNM condition during the read access but PWD problem remain exist.

In particularly, half-selected or PWD poses a critical limitation to the low-voltage or energy constrained SRAM bitcell designs. Therefore, a viable alternative (non-interleaved) mainly suitable for isolated read-port bitcells, is shown in Fig. 1.20. The non-interleaved SRAM array structure requires extra hardware and more complex layouts. For example, each word has its own word-line drivers. However, it eliminates half-selection condition during both read and write operations.

1.5 SRAM Bitcell Topologies

Standard 6T SRAM bitcell topology as shown in Fig. 1.21 has been widely used in the implementation of cache memory in high performance microprocessors and on-chip caches in SoC products. Recently, several SRAM bitcell topologies that have been proposed to achieve different objectives such as minimum bitcell area, low static and dynamic power dissipation, improved performance and better parametric yield in terms of SNM and WAM. However, other techniques such as boosting the supply voltage, read and write assist circuitries in SRAMs have also been proposed to achieve more stable data retention during read operations [53, 66]. The prime concern in SRAM bitcell design is a trade-off among these design metrics. For example, in sub-threshold SRAMs, noise margin (robustness) is the key design parameter and not speed [111, 112]. Therefore, on the basis of their robustness these bitcell topologies are broadly divided into two categories: (1) non-isolated read-port SRAM bitcell topologies (less robust), and (2) isolated read-port SRAM bitcell topologies (highly robust).

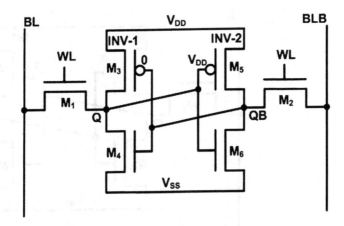

Fig. 1.21 Schematic diagram of a standard six-transistor (6T) SRAM bitcell

1.5.1 Non-isolated Read-Port SRAM Bitcell Topologies

This category of bitcell topologies are less robust because of their poor read SNM. Both read and write operations in these bitcells are performed via same pass-gate device(s) or in other words read and write ports are non-isolated. The main implication behind this kind of arrangement is tuning of bitcell ratio (β) to achieve the non-destructive read (sufficient read SNM, atleast 10–15% of V_{DD}) and successful write operations (enough WAM) simultaneously. As both the operations are conflicting in nature, hence, properly sized devices are highly desirable. A classical example of this category is a standard 6T SRAM bitcell, as shown in Fig. 1.21, where, both read and write operations are performed via pass gate devices, M_1 and M_2. Some of the recent developments in this category are as follows:

1.5.1.1 Five-Transistor (5T) SRAM Bitcell Topology

A high density low leakage five-transistor (5T) SRAM bitcell, as shown in Fig. 1.22 [20], has only one bitline and both read and write operations are performed via single pass-gate device M_1. Writing of '1' or '0' into the 5T bitcell is performed by driving the bitline to V_{DD} or V_{SS} respectively, while the wordline (WL) is asserted at V_{DD}. Sufficient WAM and read-SNM of the bitcell are ensured by cleverly sizing the transistors. This design has 15–21% smaller area for different processes, 75% lower bitline leakage, and a read/write performance comparable to standard 6T bitcell. Major drawbacks of this design are: poor static noise margins (SNM and WNM) and an on-chip DC–DC converter to generate a pre-charge voltage V_{PC} for bitline. A non-destructive read operation requires a pre-charge voltage V_{PC}, where $V_{SS} < V_{PC} < V_{DD}$. This is in contrast to the conventional 6T SRAM bitlines, which are precharged at V_{DD} before a read and write operation. The possible bitline pre-charge

Fig. 1.22 Schematic diagram of a high density low leakage five-transistor (5T) SRAM bitcell [20]

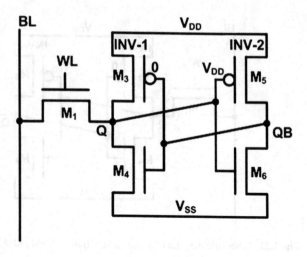

voltage V_{PC} levels (340–860 mV worst case). Therefore, in this work, a V_{PC} of 600 mV has been chosen as the bitline pre-charge voltage level for a 0.18 μm CMOS Technology.

1.5.1.2 Transmission Gate Based Six-Transistor (6T) SRAM Bitcell

In this design [120] a transmission gate (TG) is used for accessing and transferring the data during read and write cycle via single ended bitline (data line). Employing a TG for reading from and writing into a memory bitcell is dangerous because it conducts perfectly for '0' and '1' in both the directions. In other words, a read cycle can easily flip the bitcell content, because of direct and strong intervention of read-current to data storage node Q, as shown Fig. 1.23. Hence, the use of TG drastically reduces the read SNM, to even worse than the standard 6T SRAM bitcell for an iso-area topology. In TG 6T bitcell, all the devices have to be sized too large to achieve adequate read SNM, making the bitcell size more than twice the standard 6T bitcell, thus defeating the benefits of technology scaling. The write-ability of the bitcell is ensured by sharing a header and footer per column. However, sharing a header and footer per column will affect the stability of the other (non-accessed) bitcells connected to the same bitline during the write cycle. Because at the onset of a write-cycle (when signal *wr_en_* is activated), the footer will slow down the re-generative action of all the bitcells (or cross-coupled inverters) those sharing the same column for improving the write-ability of an accessed bitcell. The measurement results from a 2 Kb SRAM test-chip fabricated in 0.13 μm bulk CMOS show a 64% energy saving as compared to a multiplexor (MUX) based memory [111]. While this bitcell topology occupies more than 42% of area overhead as compared to standard 6T bitcell that fails to operate below 720 mV but it manages to operate at sub-200 mV V_{DD}. Another major drawback with this design is that the increased leakage current from the bitline (BL) limits the number of bitcells per bitline to 16.

Fig. 1.23 Schematic diagram of a transmission gate based access six-transistor (TG 6T) SRAM bitcell topology [120]

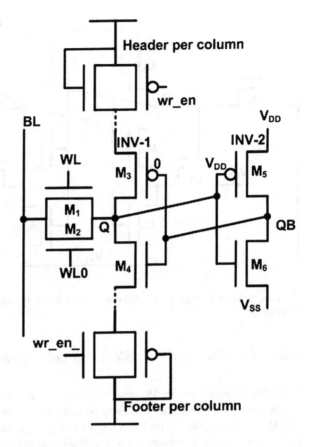

1.5.1.3 Ten-Transistor (10T) Schmitt Trigger SRAM Bitcell

Schmitt Trigger (ST) based 10-transistor SRAM bitcell topology, as shown in Fig. 1.24 [63] focuses on making the inverter pair – a basic unit of the bitcell – robust. Inverter characteristics were improved by using a modified Schmitt trigger configuration. This modified configuration is used to form the inverter pair of the bitcell. It increases or decreases the switching threshold voltage of an inverter depending on the direction of the input transition, hence, providing an enhanced ideal inverter characteristic, which effectively augments the read and hold SNMs. As a result, better read stability and write-ability in the 10T ST bitcell for subthreshold operation that comes with almost double the area overhead, as compare to standard 6T bitcell. Read and hold SNM of this design are $1.56\times$ and $2.3\times$ better than the standard 6T SRAM bitcell at $V_{DD} = 0.4$ V, respectively. More power saving in this design at lower V_{DD}, approximately 18% and 50% saving in leakage and dynamic power, respectively. These results are based on 130 nm process.

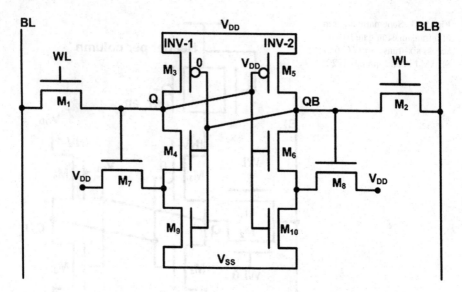

Fig. 1.24 Schematic diagram of a Schmitt Trigger based ten-transistor (10T) SRAM bitcell topology [63]

1.5.2   Isolated Read-Port SRAM Bitcell Topologies

The isolated read-port SRAM bitcell topologies also known as read-SNM free SRAM bitcell topologies have recently attracted lot of attention [18, 23–25, 73, 101, 103, 108, 111], specifically for lower supply voltage V_{DD} or sub-threshold operation. Reduction in supply voltage V_{DD} drastically reduces the SRAM bitcell noise margins and increases susceptibility to process variation. Therefore, researchers believe that it is necessary to move from standard six-transistor (6T) to eight-transistor (8T) or ten-transistor (10T) register file (1-read/1-write) type of bitcells topologies, to cope with poor noise margins or process variability problems when chips operate at lower voltages. Increasing the number of transistors to provide a separate read-port will yield extra silicon overhead and inability of scaling the SRAM bitcell for future generations. In read-SNM-free SRAM bitcells, both read and write operations are performed via separate pass-gate devices, while in the standard 6T bitcell the same pass-gate devices are used for both read and write operations. Hence, optimization of the devices size in read-SNM free SRAM bitcells is easier for a target performance and stability parameters. Mostly cited SRAM bitcell topologies base on isolated read-port are as follows.

1.5.2.1   Eight-Transistor (8T) SRAM Bitcell Topology

Figure 1.25 shows the read SNM free 8T bitcell [23, 25, 101, 103, 108], a register file type of SRAM bitcell topology, which has separate read and write ports.

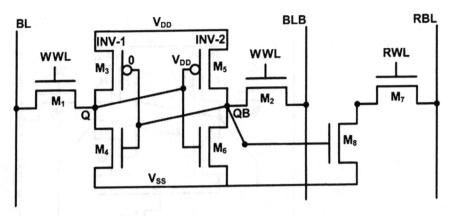

Fig. 1.25 Schematic diagram of an isolated read-port (or read SNM free) eight-transistor (8T) SRAM bitcell topology [23]

These separate read and write ports are controlled by separate read (RWL) and write (WWL) wordlines and used for accessing the bitcell during read and write cycles, respectively. In 8T bitcell topology, read and write operations of a standard 6T SRAM bitcell shown in Fig. 1.21 are de-coupled by creating an isolated read-port or read buffer (comprised of two transistors, M_7 and M_8). De-coupling of read and write operations yields a non-destructive read operation or SNM-free read stability. The interdependence between stability and read-current is overcome, while dependence between density and read-current remains there. An additional leakage current path is introduced by the separate read-port which increases the leakage current as compared to standard 6T bitcell. Therefore, an increased area overhead and leakage power make this design rather unattractive, since leakage power is a critical SRAM design metric, particularly for highly energy constrained applications.

The read bitline leakage current problem in the 8T bitcell is similar to the problem in the standard 6T bitcell, except that the leakage currents from the un-accessed bitcells and from the accessed bitcell affect the same node, RBL. So, the leakage currents can pull down RBL regardless of the accessed bitcells state. In [108] the bitline leakage current from the un-accessed bitcells is managed by adding a buffer-footer, shared by the all bitcells in that word.

1.5.2.2   Nine-Transistor (9T) SRAM Bitcell Topology

Standard 6T bitcell along with three extra transistors were employed in nine-transistor (9T) SRAM bitcell [73], to bypass read-current from the data storage nodes, as shown in Fig. 1.26. This arrangement yields a non-destructive read operation or SNM-free read stability. Results based on 65 nm process reveals that the read SNM is approximately 2× better than the standard 6T SRAM bitcell.

Fig. 1.26 Schematic diagram
of an isolated read-port
nine-transistor (9T) SRAM
bitcell topology [73]

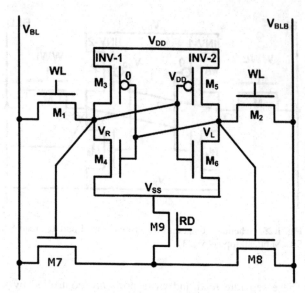

It also claims the leakage power saving of about 23% better than the standard 6T
SRAM cell. However, it leads to 38% extra area overhead and a complex layout.
Thin cell layout structure does not fit in this design and introduces jogs in the poly.

1.5.2.3   Ten-Transistor (10T) SRAM Bitcell Topology

In the 10T bitcell as shown in Fig. 1.27 [18], a separate read-port comprised of
4-transistors was used, while write access mechanism and basic data storage unit
are similar to standard 6T bitcell. This bitcell also offers the same benefits as the
8T bitcell, such as a non-destructive read operation and ability to operate at ultra
low voltages. But the 8T bitcell does not addresses the problem of read bitline
leakage current, which degrades the ability to read data correctly. In particularly, the
problem with the isolated read-port 8T cell is analogous to that with the standard
(non-isolated read-port) 6T bitcell discussed. The only difference here is that the
leakage currents from the un-accessed bitcells sharing the same read bit-line, RBL,
affect the same node as the read-current from the accessed bitcell. As a result,
the aggregated leakage current, which depends on the data stored in all of the
unaccessed bitcells, can pull-down RBL even if the accessed bitcell based on its
stored value should not do so. This problem is referred as an erroneous read. This
design has good performance at lower V_{DD}. For example, at 400 mV it can operate
at 475 kHz with small power consumption of 3.28 uW. It also has good reliability,
that is, it can operate without read error at 27°C and 320 mV, while it can have 256
bitcells per bitline.

The erroneous read problem caused by the bitline leakage current from the un-
accessed bitcells is managed by this 10T bitcell by providing two extra transistors

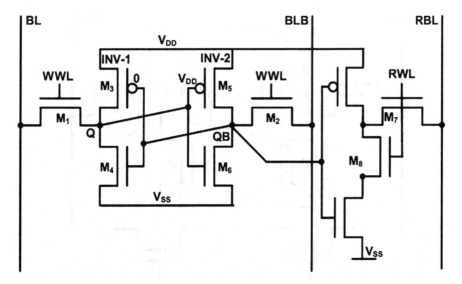

Fig. 1.27 Ultra-low voltage subthreshold ten-transistor (10T) SRAM bitcell topology [18]

in the read-port. These additional transistors help to cut-off the leakage current path from RBL when RWL is low and makes it independent of the data storage nodes content. In [58] the Reverse Short Channel Effect (RSCE) in subthreshold region was utilized to improve the write-ability and data-dependent bitline leakage current. The RSCE reduces the V_{TH} and exponentially increases the device current. Therefore, in this design, access devices utilize the RSCE to increase the write-ability. However, the RSCE would not be effective and may lead to a poor read-stability if the same access devices are used for read operation.

However, this list is not complete and there are several other potential bitcell topologies exist in the literature some of them are from [4, 44, 84, 103, 113]. In non-isolated read-port bitcell designs a non-destructive read or a successful write operation is ensured by either increasing the bitcell size or an additional circuitry or both. However, isolated read-port SRAM bitcell topologies also lead to extra area overhead but yield higher read SNM and supply voltage scaling for low power applications. Hence, a bitcell topology that yields the benefit of higher read SNM and supply voltage scaling for subthreshold operation with minimum area overhead is highly desirable.

1.5.3 Low-Leakage Asymmetric SRAM Bitcell

Standard 6T SRAM bitcell use a symmetric configuration of six transistors with identical leakage and threshold voltage (V_{TH}) characteristics. Reduction in leakage can be achieved by using higher V_{TH} transistors, however, employing all the high

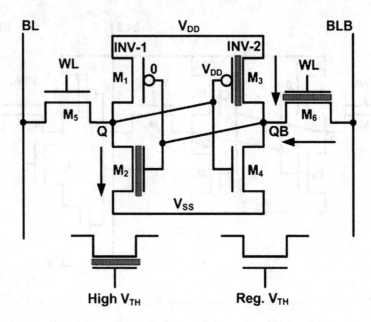

Fig. 1.28 Basic asymmetric six-transistor (6T) SRAM bitcell for low-leakage current when holding a '0'. High V_{TH} transistors are employed where leakage dissipation is higher

V_{TH} transistors will degrade the bitcell performance significantly and even up to an unacceptable level. Dual-V_{TH} technology leads to a difference of approximately $10\times$ between the leakage current of high V_{TH} and regular V_{TH} transistors. The use of dual-V_{TH} technology introduces asymmetry in the SRAM bitcell design because of different threshold transistors are employed. Therefore, SRAM bitcells in which leakage is reduced by employing dual-V_{TH} technology are called low-leakage asymmetric SRAM bitcells. Leakage current can also be reduced by increasing the length of a transistor, but it increases the area of the transistor at least 100-fold. Therefore, dual-V_{TH} technology based SRAM bitcells have good potential of saving leakage power. These bitcells exhibit asymmetric leakage and access behaviour [5, 8].

In low-leakage asymmetric SRAM bitcells, selected transistors are "weakened", or in other words V_{TH} of selected transistors is raised to reduce the leakage current, however, it can also be achieved by appropriate sizing of the transistors when the bitcell is storing a zero. In order to identify the leaky transistors, most of the bits in caches are zeros for both instruction and data streams fact is exploited. It has been shown that this behaviour persists for a variety of programs under different assumptions about the cache sizes, organization and instruction set architecture [86]. Therefore, "0" storing state is considered and weaken the only transistors necessary to drastically reduce the leakage current. In general bitcell spend most of time in the inactive state. In this state, most of the leakage is dissipated by the transistors that are in off state and that have a voltage difference across their drain and source terminals. When the bitcell storing a "0", as shown in Fig. 1.28, the leaky transistors are M_2, M_3 and M_6, therefore, these transistors are made of high threshold voltage. Similarly, if the bitcell was storing a "1", then transistors M_1, M_4 and M_5 would dissipate leakage power.

In this asymmetric bitcell, as shown in Fig. 1.28, read access time is degraded due to M_2's and M_6's higher threshold voltage which causes bitline discharge time longer. The HSPICE simulated results of this asymmetric SRAM bitcell based on 130 nm process exhibit the same leakage as the regular threshold voltage bitcell when holding a logic "1", when it holding a logic "0" leakage current is reduced by $70\times$. Therefore, there are many possible combinations to form a asymmetric SRAM bitcell keeping other parameters in mind such as saving in leakage power, stability and performance.

1.6 Summary

The importance of Static Random Access Memory (SRAM) in different processors and system-on-chip (SoC) products has fuelled the need of innovation in this area. The impact of CMOS technology scaling reduces the stability and noise margins of conventional SRAM design, as a result, alternate SRAM designs need to explore for the future technology generations. Study of different SRAM bitcells showed that the isolated read-port SRAM bitcells have better read stability and process variation tolerance as compared to non-isolated read-port SRAM bitcells. However, complex layout and extra silicon area overhead are the major limiting factor in the isolated read-port SRAM bitcells. Apart from SRAM bitcell topologies, SRAM array architectures are equally responsible for overall performance and reliability of a cache module. It is showed that the non-interleaved array architecture is more favourable for the isolated read-port SRAM bitcells and it helps in sharing the peripheral circuitries as well as minimizes the partial write disturbance.

Chapter 2
Design Metrics of SRAM Bitcell

2.1 Standard 6T SRAM Bitcell: An Overview

This section gives a brief overview of the standard six-transistor (6T) SRAM bitcell and its operation. A standard 6T SRAM bitcell consists of two identical CMOS inverters (INV-1 and INV-2) connected in a positive feedback loop. It forms a basic unit, that is, a flip-flop or latch to create a bi-stable circuit allowing the storage of one-bit of information, either '1' or '0'. The internal nodes (Q and QB) of the bitcell always contain complementary values, as shown in Fig. 2.1. The cross-coupled inverter pair itself consists of two PMOS pull-up devices (M_3 and M_5) and two NMOS pull-down devices (M_4 and M_6). Two NMOS pass-gate or access devices (M_1 and M_2), which are controlled by the wordline (WL), serve as switches between the inverter pair and the complementary pair of bitlines (BL, BLB) also called datalines, used to read in or write to the bitcell, as shown in Fig. 2.1. The data in SRAM bitcell is stored as long as the power is maintained to the bitcell. The cross coupled inverter pair can be in one of the two stable states of an SRAM bitcell, which corresponds to the data stored '1' and '0', as shown in Fig. 2.2a, b, respectively. The basic operations of a bitcell as a storage device are reading or writing new data to the bitcell.

The read and write operations in a standard 6T SRAM bitcell are performed in the following ways.

2.1.1 Read Operation

Without loss of generality, it is assumed that the internal data storage nodes Q and QB are at '0' and '1', respectively, which correspond to Fig. 2.2b. To read the bitcell contents, the following sequence of steps are performed:

Fig. 2.1 Schematic diagram
of a standard 6T SRAM
bitcell. Internal data storage
node Q at '0' and QB at '1'

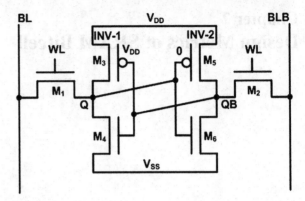

Fig. 2.2 Standard 6T SRAM
bitcell under different stable
states. (**a**) SRAM bitcell '1'
stored. (**b**) SRAM bitcell '0'
stored

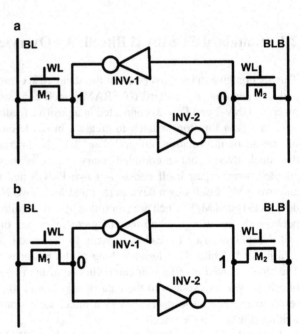

- Conventionally to read a bitcell, the bitlines (BL and BLB) are precharged to the supply voltage (V_{DD}). In some SRAM designs these bitlines are precharged to intermediate level of 0 and V_{DD}.
- The wordline (WL) is asserted to high.
- Rise the WL from '0' to '1', result, one of the bitcell sides (node) stores the logical '0'; that side of the bitline is discharged through the pass-gate and pull-down transistors. In standard 6T, as shown in Fig. 2.1, devices M_1 and M_4 discharges the precharged bitlines BL.

    - If BLB goes to low (or discharges), then the bitcell holds a logic '1' value, which correspond to Fig. 2.2a.

- If BL goes to low (or discharges), then the bitcell holds a logic '0' value, which correspond to Fig. 2.2b.

- Depending upon whether the bitline BL or BLB is discharged, the bitcell is read as a logical '1' or '0'. A sense amplifier converts the differential signal exists on BL and BLB to a logic-level output.
- De-assert the wordline (WL) back to 0.

During a read operation, the internal node (say Q) of the bitcell that holds a logical '0' will pull its bitline (BL) low through the pass-gate transistor, M_1 and pull-down transistor, M_4. It is important that the low internal node (Q) should not rise above the trip-point (switching threshold voltage) of the inverter INV-2, as shown in Fig. 2.1, to avoid a destructive read operation. A destructive read operation can be prevented by ensuring a large enough bitcell ratio (β), in other words, pull-down transistors (M_4 and M_6) must be stronger than the access transistors (M_1 and M_2). For a symmetric bitcell, bitcell ratio (β) is defined as:

$$\beta = \frac{W_4/L_4}{W_1/L_1} = \frac{W_6/L_6}{W_2/L_2} \qquad (2.1)$$

In general, the bitcell ratio can be varied from 1.25 to 2.5 depending on the target application and desired static noise margin (SNM). A larger bitcell ratio makes the bitcell robust and provides higher SNM and read current I_{read} (and hence – the speed), at the expense of increased silicon overhead and leakage current. A smaller bitcell ratio, whilst maintaining an adequate speed and noise margin, makes the bitcell compact for a high density cache design but more vulnerable to process variation induced failures.

2.1.2 Read SNM Measurement

The best measure to quantify the stability of an SRAM bitcell during the read cycle and in hold state is the Static Noise Margin (SNM). The SNM is defined as the maximum amount of DC noise (V_N) that can be tolerated by the cross-coupled inverter pair such that the bitcell retains its data [94]. The read SNM is extracted from the read voltage transfer characteristics (VTC). The read VTC can be measured by sweeping the voltage at the data storage node Q (or QB) with both bitlines (BL, BLB) and wordline (WL) biased at V_{DD} while monitoring the node voltage at QB (or Q).

Figure 2.3 shows a conceptual schematic diagram of the bitcell for the SNM definition. The bitlines (BL and BLB) induced noise is modeled with the two DC voltage noise sources (V_N), and they are introduced at each of the internal nodes

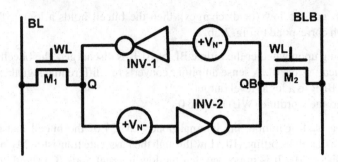

Fig. 2.3 Schematic diagram of a standard 6T SRAM bitcell showing the worst case polarity noise sources for modelling the static noise margin

in the bitcell in worst case polarity. As V_N increases the stability of the bitcell changes. The bitcell stability during active operation (read cycle) represents a more significant limitation to SRAM operation than the hold state. Specifically, at the onset of a read cycle, the wordline is activated and bitlines are precharged to V_{DD}. The internal storage node of the bitcell that represents a logic bit value '0' gets pulled upward through the access transistor due to voltage dividing phenomenon across the access transistors (M_1 and M_2) and pull down transistors (M_4 and M_6). This increase in voltage severely degrades the SNM during the read operation (read SNM), which is primarily determined by the ratio of the pull down transistor to access transistor, known as bitcell ratio.

As shown in Fig. 2.3, the noise sources are included with worst-case polarity to model the read SNM. When a worst case static noise is applied, this causes the inverse voltage transfer characteristics (VTC^{-1}) for INV-2 to move downward and the VTC of INV-1 to move to the right direction. Once both curves move by the SNM value, the meta-stable point coincides with one of the stable points and curves meet at only two points, as shown with dotted lines in Fig. 2.4. Any further noise applied to the VTC's has only one intersect point and the bitcell content flips. The voltage transfer characteristics (VTC's) and inverse VTC's (VTC^{-1}) of the two cross-coupled inverters during the read cycle are shown Fig. 2.4 for determining the worst case read SNM graphically. The SNM is estimated graphically as the length of a side of the largest square that can be embedded inside the lobes of the butterfly curve [94].

Apart from the read SNM obtained from the VTCs, the same phenomena can also be observed during transient read operation. Figure 2.5 shows the 6T SRAM bitcell read operation. The bitlines BL and BLB initially floated high. Without loss of generality, it was assumed that node Q is initially at '0' and thus QB is initially at '1'. When both the bitlines are pre-charged and word-line WL is enabled, BL should be pulled down through transistor M_1 and M_4, since both M_1 and M_4 form

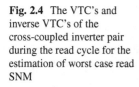

Fig. 2.4 The VTC's and
inverse VTC's of the
cross-coupled inverter pair
during the read cycle for the
estimation of worst case read
SNM

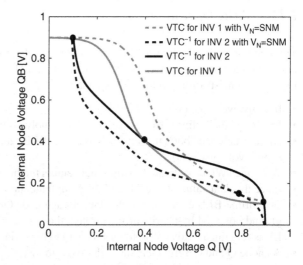

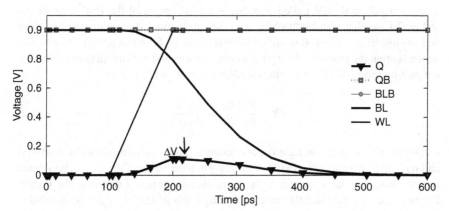

Fig. 2.5 The read operation of a standard 6T SRAM bitcell showing the stability constraint

a potential divider and raise node Q potential to ΔV. If node Q potential rises above the trip-point of inverter 2, as shown in Fig. 2.1, than there will be a read failure. This stability constraint is referred to as a 'read destructive operation'. In this case, wordline is kept high for longer period as a result bitline (BL) is pulled to ground, however, wordline pulse width is usually kept enough smaller that can develop a small differential voltage on the bitlines (BL and BLB). This small differential voltage overcomes the offset voltage of a sense amplifier in order to latch the correct data from the bitlines.

2.1.3 Write Operation

The write operation or flipping the bitcell contents when initially assuming that the internal data storage nodes Q and QB are at '0' and '1', respectively, as shown in Fig. 2.2b, consists of the following sequence of steps:

- Initially, wordline (WL) = 0.
- Precharge the bitlines (BL and BLB) to the supply voltage (V_{DD}).
- After prechrage, both the bitlines (BL and BLB) are disconnected from the supply voltage (V_{DD}).
- Wordline (WL) is activated to high (data enters the bitcell during this step).
- Place the data value on the BL and the complementary data value on BLB.
- The bitline BLB connected to the data storage node QB via M_2, is driven to the ground potential by a write driver through the M_2 pass-gate transistor, while the BL is remained held at V_{DD} to pull node Q to high via M_1 pass-gate transistor.
- As node Q and QB flip their states de-assert the wordline (WL) back to 0.

As a result, node QB pulled well below the trip-point of the INV-1, as shown in Fig. 2.1, so that, the feedback loop in the cross-coupled inverter starts to work and the bitcell is flipped. It is important that a successful write operation primarily depends upon the properly sized pass-gate transistors and pull-up transistors. Hence, the pull-up ratio (PR) for a symmetric SRAM bitcell is defined as:

$$PR = \frac{W_3/L_3}{W_1/L_1} = \frac{W_5/L_5}{W_2/L_2} \qquad (2.2)$$

During write operation fight occurs between pass-gate transistors (M_1 and M_2) and pull-up transistors (M_3 and M_5) mainly when the value of the data to be written into the bitcell is the opposite of the value that is currently stored in the bitcell. For instance, assumed that the bitcell currently holds a logic '1' and it is intended to write a logic '0' into the bitcell. Then, the situation correspond the following biasing conditions; BLB, WL and bitcell supply voltage is biased at V_{DD} and BL is biased at V_{SS} in Fig. 2.2a. For a successful write operation, node QB must be pulled above the trip point of the inverter "INV1" and node Q must be pulled down below the trip-point of the inverter "INV2". In this mode, fight occurs between M_1 and M_3. Since M_1 will try to bring node Q to down while M_3 will try to keep node Q to high. Therefore, M_1 must win the fight for a successful write operation. Once an inverter switches, it creates a self-reinforcing positive feedback action that continue until the bitcell has been fully placed into a new stable state.

A successful write operation can be guaranteed by choosing a lower PR value (generally, PR = 1), that can be achieved by employing the wider or stronger pass-gate transistors (M_1 and M_2) instead of pull-up transistors (M_3 and M_5). However, increasing the width of the pass-gate transistors threatens the stability of the bitcell during the read cycle, or in other words reduces the read SNM of the bitcell.

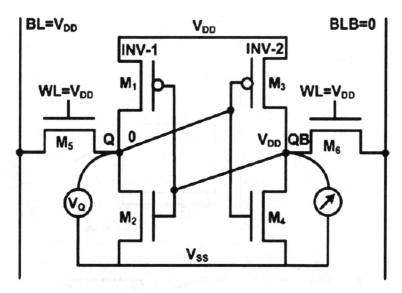

Fig. 2.6 Experimental set-up for extracting the write static noise margin

2.1.4   Write SNM Measurement

Figure 2.6 shows the experimental set-up for measurement of write static noise margin (WSNM). During write operation M_1 (or M_3) and M_5 (or M_6) form a resistive voltage divider for the falling BL (or BLB) and storage node Q (or QB). If the voltage divider pulls Q below the trip-point of inverter 2 (INV-2), a successful write operation occurs. The write-ability of a SRAM bitcell can be gauged by the write SNM [12, 13]. The write SNM is extracted by a combination of read voltage transfer characteristics (VTC) and the write VTC. The read VTC is measured by sweeping the voltage at the storage node QB while monitoring the node voltage at Q. The write VTC is measured by sweeping the voltage at the storage node Q with BL and WL biased at V_{DD} and BLB is biased at V_{SS} while monitoring the voltage at node QB. The WSNM can be quantified by the side of the smallest square fitted inside the read and write VTCs, as shown in Fig. 2.7. When WSNM fall negative, the read VTC and write VTC intersects each other suggesting the inability to write into the SRAM bitcell.

The write-ability of an SRAM bitcell can also be characterized by the write trip-point-voltage, which is defined as the maximum amount of voltage needed on the bitline to flip the bitcell content [36, 40]. The Write Static Noise Margin (WSNM) is measured through the write-trip point defined as the difference between V_{DD} and the maximum bitline voltage required to flip the data storage nodes Q and QB, as shown in Fig. 2.8.

Fig. 2.7 Measurement of
write static noise margin
(WSNM) of an SRAM bitcell
obtained from read and write
VTCs

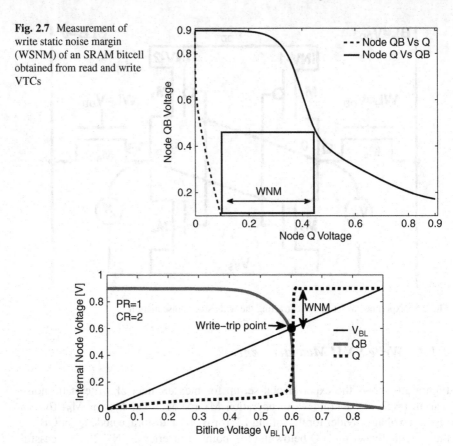

Fig. 2.8 The write noise margin (WNM) of an SRAM bitcell defined as write-trip point

2.1.5   Relationships Between Transistor Drive Strengths

A concern associated with the read operation is that both BL and BLB are kept high
at the beginning of the read operation must not corrupt (un-intended flip) the value
stored in the bitcell. In order to avoid the un-intended flip of the value, it is desirable
to keep the voltage at the internal node which has a stored value of '0' from rising
above the trip point of the inverter, i.e. more of the voltage drop between bitline
and ground should occur across the access transistors (M_1 and M_2) than across the
pull-down transistors (M_4 and M_6), refer Fig. 2.1. In other words, the strength of the
access transistors should be less than the strength of the pull-down transistors for a
non-destructive read operation.

Similarly, for a successful write operation, it is desirable to bring down the voltage of the internal data storage node Q (or QB) which has a stored value '1' below
the trip point of the inverter, INV-2 (or INV-1), refer Fig. 2.1. Therefore, access

transistors (M_1 and M_2) must be stronger than the pull-up transistors (M_3 and M_5) for a successful write operation. Combining these constraints, yield the following relation:

strength(PMOS Pull-up) < strength(NMOS Access) < strength(NMOS Pull-down).

2.2 Other SRAM Bitcell Stability Metrics

In the recent past there has been substantial efforts were made to understand and model the stability of an SRAM bitcell. Many analytical models have been developed for the static noise margin (SNM) of an SRAM bitcell to optimize the bitcell design, to predict the effect of parameter variations on the SNM [95] and to access the impact of intrinsic parameter variations on the SRAM bitcell stability [10]. Some of the recent methods for measurement of SRAM stability are N-curve and bitline measurement techniques. SRAM stability metrics can broadly be classified into two categories: static and dynamic. In static SRAM stability metrics, a DC voltage is applied or swapped to estimate how much DC noise an SRAM bitcell can tolerate. The static stability metrics are further divided into two categories: conventional and large scale metrics, as shown in Fig. 2.9. Conventional metrics are based on butterfly curves and N-curves. These metrics require access of internal nodes of an SRAM bitcell for measurement of stability and provides limited data for stability analysis. Access of internal nodes of all the SRAM bitcells is not feasible because of metal spacing constraints and area overhead associated to provide switch array. On the other hand, large scale metrics do not require access of internal data storage nodes for stability analysis, therefore, these metrics are suitable for dense large scale SRAM designs. In theses metrics, measurement of stability is done by accessing the bitlines (BL or BLB), wordline (WL), and bitcell supply (V_{CELL}).

2.2.1 N-Curve Stability Metrics

The stability of an SRAM bitcell is commonly defined by the *SNM* as a maximum value of DC noise voltage that can be tolerated without changing the internal storage node state [1, 41, 103]. A successful data retention during hold and functional operations read and write are determined by hold *SNM*, read *SNM* and write trip voltage, respectively. These three metrics are widely used for design and performance analysis of SRAM bitcell but none of the metrics carry the *current flow* information which is having extensive importance. For example, in hold state the hold *SNM* is highly dependent on the driving capability of the pull down NMOS transistors, whereas read *SNM* is strongly dependent on the driving capability of the NMOS access and pull down transistors.

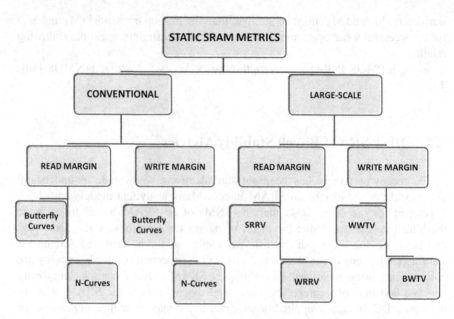

Fig. 2.9 Classification of static stability metrics of a standard 6T SRAM bitcell

The major drawbacks of *SNM* metric obtained from butterfly curves are as follows:

- The ideal voltage transfer characteristic (VTC) obtained from the butterfly curve delimits to a maximum $0.5V_{DD}$,
- Inability to measure it with a automatic inline tester, due to the fact that after measuring the butterfly curves of the bitcell, SNM still has to be derived by mathematical manipulations of the measured data,
- Inability to generate statistical information of SRAM failures, due to indirect availability of SNM, and
- Separate analysis is required for read and write stability measurement,
- It does not provide *current flow* information which is equally important for stability analysis.

An alternative approach for stability analysis that satisfies the above requirements is the use of N-curve for an SRAM design [114].

Figure 2.10 shows the experimental setup for extracting the N-curve of a standard 6T SRAM bitcell. The same setup can also be extended for other SRAM bitcell topologies. At the beginning of read access both bitlines (BL and BLB) are precharged to '1' and wordline is activated to '1'. Without loss of generality, it is assumed that the internal storage nodes QB and Q are at '1' and '0', respectively. A voltage sweep V_{in} from 0 to V_{DD} is applied at the node Q and corresponding current supplied by the sweep voltage source V_{in} is measured as I_{in}. The resulting relationship is plotted between V_{in} on x-axis and I_{in} on y-axis is

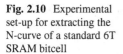

Fig. 2.10 Experimental set-up for extracting the N-curve of a standard 6T SRAM bitcell

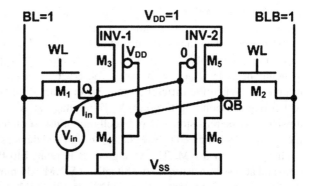

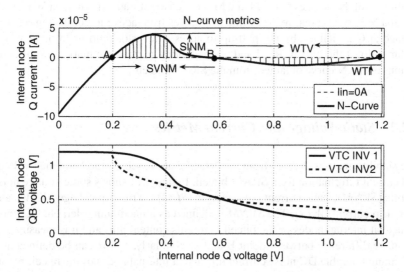

Fig. 2.11 Extracted N-curve and corresponding butterfly curves for read operation of a standard 6T SRAM bitcell

called the N-curve, as shown in Fig. 2.11. As an attractive approach, N-curve contains information for both read stability and write stability. There is no need of mathematical manipulation on the measured data as N-curve directly provides the functional information of SRAM bitcell. N-curve contains information for both voltage and current. Thus, allowing a complete functional analysis of the SRAM bitcell stability for both read and write operations with only one N-curve. However, major limitations of the N-curve based stability analysis is that it also requires the access of internal nodes of the SRAM bitcell similar to butterfly curves.

The extracted N-curve has three intersection points A, B, and C; point A and C correspond to stable state points while point B is a meta-stable point, these points corresponds to the butterfly curves plotted below the N-curve, as shown in Fig. 2.11. At these points current supplied by the sweep voltage source V_{in} is zero. At the beginning, when both V_{in} and node Q at 0 V, the access transistor M1 and transistor M4 are in velocity saturation and linear region, respectively. Therefore, drain current of M1 is larger than the drain current of M4. Thus, the difference of these currents according to Kirchoff's current law, I_{in} flows into the sweep voltage source in order to maintain node Q at 0 V. It can be observed in negative direction from origin to point A. When the difference of these currents is equal to 0 A (i.e. $I_{in} = 0$ A), which is corresponding to point A. The voltage at A is determined by the pull-down to access transistor ratio or bitcell ratio. Further increase in sweep voltage V_{in}, increases I_{in} as indicated by the change in sign and devices operation region remain unchanged up to SINM. The voltage at B is related to the pull-down to pull-up ratio and access transistors of the bitcell. At SINM, M4 moves from linear region to velocity saturation region. Between SINM and WTI, M3 is now active and working regions of all the devices M1, M4 and M3 moved to saturation region. At WTI, both M1 and M3 are in linear region while M4 moves from active to cut-off region. The voltage at C is defined by the pull-up to access transistor ratio or in other words pullup ratio of the bitcell. Read and write stability metrics are marked in different regions of the N-curve obtained from the SRAM bitcell.

2.2.2 Static Voltage and Current Metrics

The stability metrics derived from the N-curve are based on the combined voltage and current information for a SRAM bitcell. Figure 2.11 shows static voltage noise margin (*SVNM*), static current noise margin (*SINM*), write trip voltage (*WTV*), and write trip current (*WTI*). The *SVNM* is defined as a maximum tolerable DC noise voltage at internal nodes of the bitcell before its content flips and it is measured as a voltage difference between point B and A. Similarly, *SINM* can be defined as a maximum tolerable DC noise current injected at internal nodes of the bitcell before its content changes and it is measured as a peak current located between point A and B. These two metrics *SVNM* and *SINM* are used to characterize the bitcell read stability.

However, bitcell's write stability can be characterize the with the help of *WTV* and *WTI*. For this purpose N-curve has to be analyzed from right to left because for write operation, pulling down of precharged bit line (BLB) to ground so that the internal node QB get discharges. The *WTV* is the minimum voltage drop needed to change the internal nodes of the bitcell, which can be measured as a difference between point C and B. The *WTI* is defined as a minimum amount of the current needed to write the bitcell which can be measured as a negative current peak between point C and B as shown in Fig. 2.11. An overlap of points A and B or point B and C means the bitcell is at the edge of stability loss, as a result, destructive read operation can easily occur. Similarly, overlapping of these points may lead to failure in write operation.

2.2.3 Power Metrics

The N-curve as shown in Fig. 2.11 is used to derive the power metrics which includes both the voltage and current information for read stability and write stability. So, instead of using four metrics obtained from N-curve to analyze the stability of an SRAM bitcell, it would be easy to combine them in two power metrics, static power noise margin (*SPNM*) and wtite trip power (*WTP*) [97]. The *SPNM* is used to characterize the read stability which is measured as the area below the curve between point A and B. The *SPNM* is defined as the maximum tolerable DC noise power by the internal data storage nodes of the bitcell before its content changes. Furthermore, as the shaded part of N-curve between point A and B has formally a unit of power which is given by Eq. 2.3,

$$SPNM = \frac{1}{B-A} \sum_{n=A}^{n=B} I_{in}(n) * V_{in}(n).$$ (2.3)

The *WTP*, characterizes the write stability of a bitcell and which is measured as the area above the curve between point B and C. The *WTP* is defined as the minimum amount of power required to flip the data storage nodes and which is given by Eq. 2.4:

$$WTP = \frac{1}{C-B} \sum_{n=B}^{n=C} I_{in}(n) * V_{in}(n),$$ (2.4)

where V_{in} is the sweep voltage source and I_{in} is the current supplied by the V_{in}. The successful write in the bitcell is quantified with the help of this metric. From Fig. 2.11 it is clear that for a successful read and write operations *SPNM* should be positive (i.e. *SPNM* > 0) and *WTP* should be negative (i.e. *WTP* < 0).

2.2.4 Dependencies of SPNM and WTP

The stability of the bitcell degrades with lowering supply voltage V_{DD}, minimum bitcell size and process variability which will limit advanced technology node to operate at lower voltage due to degraded read *SNM* and reduced write margin. Read *SNM* degradation results in destructive read operation whereas reduced write margin cause unsuccessful write operation. To observe the dependency of different parameters, SPICE simulation results are presented in this section for a standard 6T SRAM bitcell, these results are based on the predictive technology model (PTM) 65 nm node.

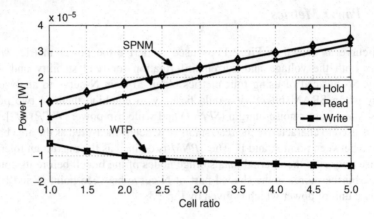

Fig. 2.12 Cell ratio dependency of SPNM and WTP at $V_{DD} = 1.2$ V

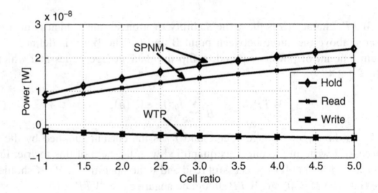

Fig. 2.13 Cell ratio dependency of SPNM and WTP at $V_{DD} = 0.3$ V

2.2.5   Dependence on the Bitcell Ratio

The stability as well as the size of the SRAM bitcell is primarily determined by the bitcell ratio, which is the defined as the ratio of pull down transistor's (W/L) to the access transistor's (W/L). Figure 2.12 shows the impact of bitcell ratio on SPNM and WTP at $V_{DD} = 1.2$ V during hold, read and write operations. As shown in Fig. 2.12, the SPNM is almost linearly increases with the bitcell ratio. The linear dependence of SPNM on bitcell ratio is because of the drain current of the pull down transistors and access transistors increases linearly with the bitcell ratio. Figure 2.13 shows that the bitcell ratio has clear impact on SPNM at subthreshold $V_{DD} = 0.3$ V during hold, read and write operations. In subthreshold, the dependence of SNM obtained from the butterfly curve has very little (unnoticeable) impact of bitcell ratio [17]. However, power metric SPNM and WTP obtained from N-curve at subthreshold $V_{DD} = 0.3$ V shows the consistent trend as it is at $V_{DD} = 1.2$ V. Hence, the

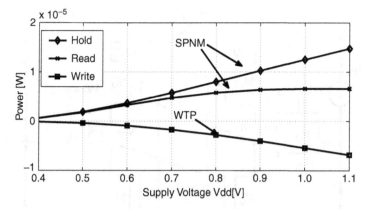

Fig. 2.14 Supply Voltage V_{DD} dependency of SPNM and WTP

proposed metrics provides better information compare to *SNM* at ultra low voltage and can be useful for stability analysis under subthreshold regime.

2.2.6 Dependence on the Supply Voltage V_{DD}

The *SNM* obtained from the VTC delimits to a maximum $0.5V_{DD}$ because of the two sides of the butterfly curve [17]. Figure 2.14 shows the dependence of power metrics *SPNM* and *WTP* on V_{DD} for a standard 6T-SRAM bitcell. The power metrics *SPNM* and *WTP* for hold, read and write operations reveals that V_{DD} scaling no longer limits the SRAM bitcell stability to the ideal value of $0.5V_{DD}$. Thus, the proposed metrics dependency on V_{DD} as shown in Fig. 2.14 will not limit the stability analysis and can be used at a very low voltage.

2.3 Bitline Measurement Design Metrics

The conventional DC read and write static noise margin (SNM) metrics presented in the previous section have some major drawbacks such as inability to measure them in dense functional SRAM arrays due to metal spacing constraints and area overhead associated to provide switch array. As a result, it produces inadequate number of data points for failure analysis of large size cache memories. In bitline measurement, SRAM read and write stability is characterized by accessing only the bitlines, wordline, and the bitcell supply voltages [38]. It increases the number of data points for failure analysis, while the SRAM array kept intact. Bitline measurement can be used for analysing the functional stability of an SRAM bitcell for read and write operations. In contrast to butterfly curve based stability

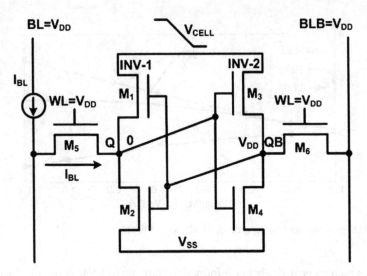

Fig. 2.15 Schematic diagram for extracting the supply read retention voltage (SRRV)

analysis, bitline measurement directly provides the functional stability without mathematically manipulating the measured data. Bitline measurement has also been previously applied to detect and isolate the faulty SRAM bitcells in cache memory array [116].

2.3.1 Read Stability Measurement

Supply Read Retention Voltage (SRRV): The read stability of an SRAM array can also be measured in terms of SRRV. For read stability measurement, both bitlines (BL and BLB) are kept floating around V_{DD} while the wordline (WL) is driven high, and the bitcell state is retained by keeping the bitcell supply sufficiently high. Therefore, SRAM bitcell read stability in functional SRAM array can be gauged by the minimum bitcell supply needed for data retention during read operation, which is referred as the supply read retention voltage (SRRV).

Figure 2.15 shows the schematic setup for extracting the SRRV. To extract the read stability using SRRV, both bitlines (BL and BLB) are tied to V_{DD} and wordline is also driven to operating voltage V_{DD} to emulate the read operation. The bitline (BL) current, I_{BL}, is monitored at the '0' storage node, while ramping down the SRAM bitcell supply (V_{CELL}) voltage. When the V_{CELL} is ramped down sufficiently low, the SRAM bitcell loses its stability for data retention and makes nodes Q and QB monostable. At this point, M_5 dominates M_2 so that node Q, originally holding '0' rises above the trip point of inverter (INV-2) and flips the bitcell state. It is also signified by the sudden drop in bitline current, I_{BL}, as shown in Fig. 2.16. Sudden

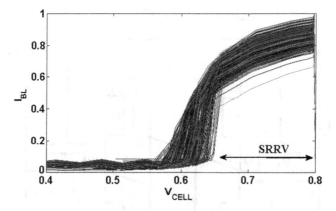

Fig. 2.16 Definition of SRRV from the measured transfer characteristics between I_{BL} and V_{CELL}. The bitcell supply (V_{CELL}) voltage is ramped below the V_{DD} causes sudden drop in bitline current

drop in I_{BL} is mainly due to rise in the potential of node Q storing '0' when it reaches to meta-stable state or bitcell state flips due to drop in V_{CELL}.

The transfer characteristics plotted between I_{BL} versus V_{CELL}, are shown Fig. 2.16 for measurement of SRRV. The SRRV of an SRAM bitcell can be defined as the difference between V_{DD} and the value of the V_{CELL} causing I_{BL} to suddenly drop, as shown in Fig. 2.16. Initially, when $V_{CELL} = V_{DD}$ (i.e. SRRV = 0), the SRAM bitcell is biased for a nominal read operation with BL, BLB, WL and V_{CELL} all are biased at V_{DD}. If the SRRV is greater than zero, it indicates that bitcell supply voltage (V_{CELL}) can be dropped below V_{DD} without disturbing the bitcell state. Therefore, SRRV represents the maximum tolerable DC noise voltage at the bitcell supply before causing the destructive read operation.

Wordline Read Retention Voltage (WRRV): During read or write operation, wordline is driven high causes read stress to SRAM bitcells having direct read access and all the unaccessed SRAM bitcells. This read stress can be further exacerbated by boosting the wordline voltage above the V_{DD}. Hence, the read stability of an SRAM bitcell can also be measured from the largest wordline boost without flipping the bitcell contents, defined as word line read retention voltage (WRRV).

Figure 2.17 shows the schematic setup for extracting the WRRV. To extract the read stability using WRRV, both bitlines (BL and BLB) are tied to V_{DD} and bitcell supply (V_{CELL}) is also driven to operating voltage V_{DD} to emulate the read operation. The bitline (BL) current, I_{BL}, is monitored at the '0' storage node, while ramping up the wordline voltage above the supply voltage (V_{DD}). However, keeping the wordline voltage below the gate-oxide breakdown voltage set by the technology. When the wordline voltage is boosted adequately high above V_{DD}, the SRAM bitcell state is flipped due to an exacerbated read stress which causes M_5 to dominate over M_2, so that node Q, originally holding '0' rises above the trip point of inverter (INV-2) and flips the bitcell state. It is also signified by the sudden drop in bitline

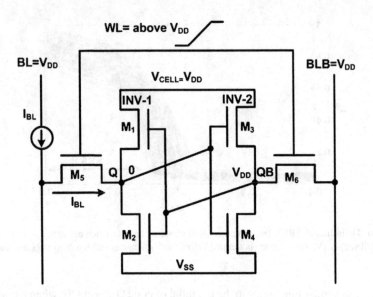

Fig. 2.17 Schematic diagram for extracting the wordline read retention voltage (WRRV)

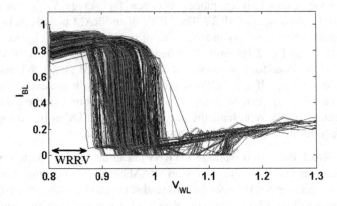

Fig. 2.18 Definition of WRRV from the measured transfer characteristics between I_{BL} and V_{WL}. The wordline voltage (V_{WL}) is ramped above the supply voltage causes sudden drop in bitline current

current, I_{BL}, as shown in Fig. 2.18. This phenomena is observed due to the potential drop between BL and node Q when rising wordline potential causes the bitcell to meta-stable state or bitcell state flips.

The transfer characteristics plotted between I_{BL} versus V_{WL}, are shown Fig. 2.18 for measurement of WRRV. The WRRV of an SRAM bitcell can be defined as the difference between V_{DD} and the boost in the wordline voltage causing I_{BL} to suddenly drop, as shown in Fig. 2.18. Initially, when $V_{WL} = V_{DD}$ (i.e. WRRV = 0), the SRAM bitcell is biased for a nominal read operation with BL, BLB, WL and

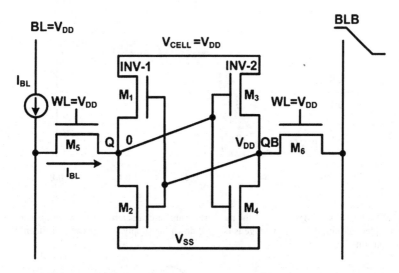

Fig. 2.19 Schematic diagram for extracting the bitline write trip voltage (BWTV)

V_{CELL} all are biased at V_{DD}. If the WRRV is greater than zero, it indicates that wordline can be boosted above V_{DD} without disturbing the bitcell state. Therefore, WRRV represents the maximum tolerable DC noise voltage at the wordline before causing the destructive read operation.

2.3.2 Writeability Measurement

Bitline Write Trip Voltage (BWTV): Figure 2.19 shows the schematic setup for extracting the BWTV. The bitcell is configured according to the new data which has to be written under write operation. In this scenario, bitline (BL) and wordline (WL) are tied to V_{DD} while complement bitline (BLB) voltage, V_{BLB} is ramped from V_{DD} to ground potential to emulate the write cycle. The writeability of an SRAM bitcell in a functional SRAM array can be gauged by the maximum bitline voltage (V_{BLB}) at the '1' storage node (QB), able to flip the bitcell state during a write cycle [33, 34, 37, 41]. To extract the BWTV of an SRAM bitcell, the bitcell supply (V_{CELL}), BL and WL are biased at V_{DD}. The bitline current (I_{BL}) is monitored while ramping down the bitline (V_{BLB}) voltage. As the V_{BLB} dropped low enough, the pass-gate M_6 overcome M_3 and the node QB is dropped below the trip-point of inverter, INV-1. As a result, a successful write operation is observed which is signified by the sudden drop in I_{BL}, as shown in Fig. 2.20. The BWTV is quantified from the transfer charactristics of I_{BL} versus V_{BLB}, as the V_{BLB} dropped low that causes the sudden fall in I_{BL}. When BWTV is zero (BWTV = 0) the SRAM bitcell is biased for a nominal write operation. If the BWTV is greater than zero,

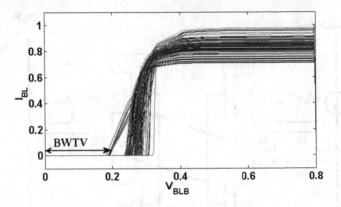

Fig. 2.20 Definition of BWTV from the measured transfer characteristics between I_{BL} and V_{BLB}. The complement bitline (V_{BLB}) voltage is ramped down the supply voltage causes sudden drop in bitline current

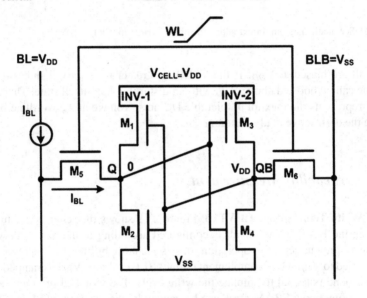

Fig. 2.21 Schematic diagram for extracting the wordline write trip voltage (WWTV)

it indicates that bitline voltage (V_{BLB}) can be dropped below V_{DD} for a successful write operation. Therefore, BWTV represents the maximum tolerable DC slack on the bitline to successfully write the bitcell.

Wordline Write Trip Voltage (WWTV): The WWTV is defined as the minimum wordline voltage needed to flip the bitcell content during a write cycle and it can be used to gauged the writeability of an SRAM bitcell in an SRAM array. Figure 2.21 shows the schematic setup for extracting the WWTV by first configuring the bitlines to write the data and then ramping the wordline. The bitcell supply (V_{CELL}) and BL

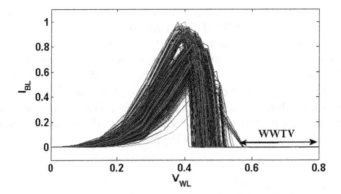

Fig. 2.22 Definition of WWTV from the measured transfer characteristics between I_{BL} and V_{WL}. The wordline (V_{WL}) voltage is ramped down the supply voltage causes sudden drop in bitline current

are biased at V_{DD} while BLB is biased and V_{SS}. The bitline current (I_{BL}) at the '0' storage node (Q) is monitored while ramping up the wordline (WL) voltage. As wordline voltage ramped high, the monitored current initially resembles the I_D-V_G charactristics of pass-gate transistor M_5. When the ramped up voltage of the WL is sufficiently high, the bitcell content flips, signified by the sudden drop in the I_{BL} magnitude as shown in Fig. 2.22.

The WWTV is quantified from the transfer charactristics of I_{BL} versus the V_{WL} voltage, as shown in Fig. 2.22. The WWTV is defined as the difference of V_{DD} and the V_{WL} voltage which causes the sudden drop in the I_{BL}. When WWTV = 0, the SRAM bitcell is biased for a nominal write operation with WL, BL and V_{CELL} are biased at V_{DD} and BLB is biased at V_{SS}. If the WWTV is greater than zero, it indicates that wordline voltage can be dropped below V_{DD} for a successful write operation. Therefore, WWTV represents the maximum tolerable DC slack on the wordline to successfully write the bitcell.

2.4 Dynamic Stability Analysis

Static stability analysis has been extensively used to characterize the SRAM bitcell failures. The most commonly used failure criteria are SNM for read failure and WNM for write failure [12, 95]. All (SNM, WNM and N-curve) static failure analysis are steady state in nature and assume the wordline (WL) is turned ON indefinitely and the bitlines (BL and BLB) are driven to V_{DD} indefinitely. These metrics are known to be optimistic in write stability and pessimistic in read stability from comparison between static and actual dynamic access [55, 82, 105, 117]. The read SNM is usually overestimates the read failure. For example, an unstable bitcell might not have enough time to flip with a finite WL pulse width for a noise magnitude larger or equal to SNM. Similarly, WNM usually underestimates the write failure. It is due to the fact that a bitcell might be too slow to be written in a finite duration of the WL pulse width even with a bitline voltage is larger or equal to WNM. It is well known that both SNM and WNM, in general, are not capable of capturing different effects such as coupling, charge sharing, transient, and other dynamic bitcell behaviour under read and write operations. Therefore, dynamic stability analysis of a SRAM bitcell is must to determine the functionality or a successful read and write operation in time domain.

2.4.1 Dynamic Read Stability

Figure 2.23a shows an SRAM bitcell under read access for different pulse widths T_A and T_B. It illustrates the importance of pulse width during read access and shows that how insufficient pulse width can lead to latch a wrong value. The schematic setup for simulation of read access is shown in Fig. 2.23a in which bitlines are precharged to V_{DD} before the read access cycle and wordline is active high during read access. In Fig. 2.23b, wordline pulse width T_A is too short, therefore, bitline capacitance is not discharged sufficiently to overcome the offset in the sense amplifier. As a result, wrong value is latched by the sense amplifier. Similarly, a successful read access operation is shown in Fig. 2.23c for a wordline pulse width T_B. The wordline pulse width T_B is good enough to discharge the bitline (BL) and causes sufficient differential voltage to overcome the offset or trigger the sense amplifier. As a result, it latches the correct value of the data stored in the bitcell. It emphasize that there should be a critical wordline pulse width, T_{access} ($T_A < T_{access} < T_B$), for which sense amplifier is on the threshold of a successful read access that is defined as the read access time. This is similar to dynamic read access failure defined in [55, 82, 105]. Thus, size of wordline pulse width has a significant role in performing the correct read operation.

Under dynamic read stress, standard 6T SRAM bitcell's bitlines (BL and BLB) are precharged to V_{DD} and wordline is activated with different pulse widths (T_A and T_B), as shown in Fig. 2.24. The wordline pulse width T_A is short enough so that the internal nodes Q and QB return back to their original levels after the wordline pulse is deasserted, as shown in Fig. 2.24b. This is a desirable feature in order to avoid the destructive read operation. However, wordline pulse width should not be shorter than the pulse width of a successful read access operation, as shown in Fig. 2.23c. A longer wordline pulse width T_B is applied to put the SRAM bitcell under heavy stress, causing the bitcell to flip to an opposite state before the wordline pulse is deasserted. It is corresponding to the destructive read failure (i.e. read upset) because bitcell's content changed during read access which is not desirable. The destructive read failure, therefore, can be defined as a bitcell is unstable if the voltage of the node storing a '0' (Q) reaches the trip point of the inverter (INV2) during the pulse width

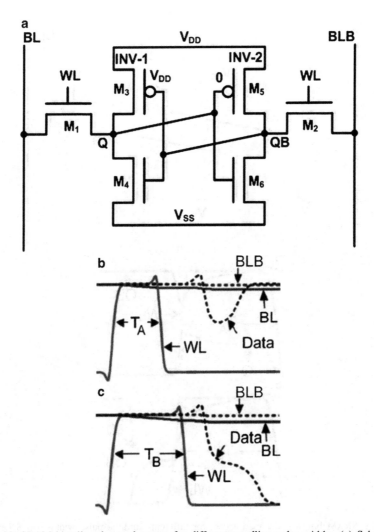

Fig. 2.23 SRAM bitcell under read access for different wordline pulse widths. (**a**) Schematic diagram of a standard 6T SRAM bitcell under read access (i.e. bitlines are precharged to V_{DD} and wordline is active) storing '0' on the left internal node Q. (**b**) Different simulated waveforms showing failed read access with wordline pulse width, T_A. The output of the sense amplifier (Data) resolves to the wrong value. (**c**) Different simulated waveforms showing successful read access with a longer wordline pulse width, T_B. The output of the sense amplifier (Data) resolves to the correct value

T_B is shown in Fig. 2.24c. The bitcell stress applied with different wordline pulse widths have significant role in determining the dynamic read stability. Therefore, there should be a critical pulse width T_{read} ($T_A < T_{read} < T_B$), for which the bitcell is on threshold of destructive read operation or read upset.

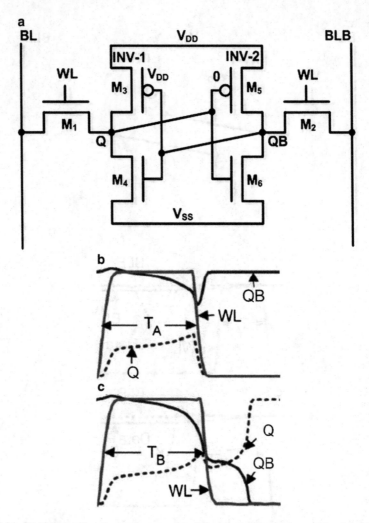

Fig. 2.24 SRAM bitcell under read access for different wordline pulse widths. (**a**) Schematic diagram of a standard 6T SRAM bitcell under dynamic read stress (i.e. both bitlines are precharged to V_{DD} and wordline is active with different pulse widths). (**b**) Different simulated waveforms showing a successful read operation with wordline pulse width, T_A. The internal data storage nodes Q and QB retained their states after read operation. (**c**) Different simulated waveforms showing destructive read operation with a longer wordline pulse width, T_B. The internal data storage nodes Q and QB accidentally flipped their states after read operation

2.4.2 Dynamic Write Stability

Dynamic write failure of a standard 6T SRAM bitcell, as shown in Fig. 2.25a can be studied under following biasing conditions that are correspond to a write operation. The bitline (BL) is charged to V_{DD} and its complementary bitline (BLB) is tied to

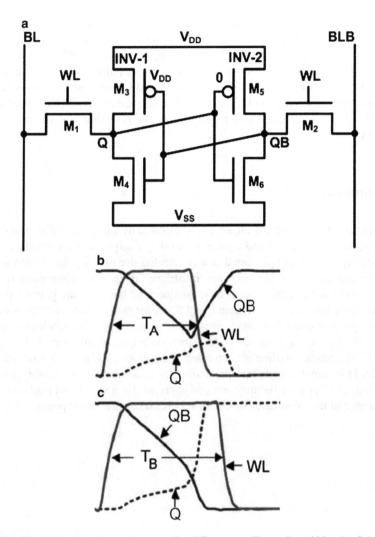

Fig. 2.25 SRAM bitcell under read access for different wordline pulse widths. (a) Schematic diagram of a standard 6T SRAM bitcell under read access (i.e. bitlines are precharged to V_{DD} and wordline is active) storing '0' on the left internal node Q. (b) Different simulated waveforms showing failed read access with wordline pulse width, T_A. The output of the sense amplifier (Data) resolves to the wrong value. (c) Different simulated waveforms showing successful read access with a longer wordline pulse width, T_B. The output of the sense amplifier (Data) resolves to the correct value

V_{SS} and the wordline (WL) is activated with different pulse widths (T_A and T_B). The wordline pulse width T_A is too short to overwrite the internal data storage nodes Q and QB of the SRAM bitcell. As a result, node Q and QB return back to their original states after the wordline pulse T_A is deasserted, as shown in Fig. 2.25b. It is corresponding to the write failure because bitcell's content are not changed during write access which is not desirable. The dynamic write failure, therefore, can be defined that if the voltage on the node initially storing '1' (QB) could not reach the trip point of the opposite inverter (INV1) during the wordline (WL) pulse then the bitcell will not successfully be written. In Fig. 2.25c, pulse width T_B is sufficient to complete the write operation, that is, node Q and QB change their states before the deassertion of wordline pulse width T_B. Therefore, there should be critical pulse width T_{write} ($T_A < T_{write} < T_B$), for which the bitcell is on threshold of a successful operation.

2.5 Summary

The functionality of a SRAM bitcell can be ensured by analysing different stability criterion, therefore, static and dynamic stability analysis techniques have been exhaustively studied and presented in a comprehensive manner. Simulation setups for different stability techniques are illustrated along with simulation results. Limitations of static stability analysis techniques and how they are pessimistic and optimistically estimates the write and read SNM for failure analysis emphasize the need of dynamic stability analysis. For large size cache memories, where butterfly curve and N-curve approaches do not provide adequate number of data point for stability analysis, bitline measurement techniques play a significant role and provides large number of failure points for statistical analysis of stability are also presented. The dynamic stability analysis approach for a successful read and write operations and the importance of wordline pulse width are also explained.

Chapter 3
Single-Ended SRAM Bitcell Design

3.1 Introduction

Those embedded systems which are particularly targeted towards low duty-cycles and portable applications, such as mobile phones or PDAs must have low energy consumption, as these systems are battery powered. In such systems, a considerable amount of power is consumed during memory accesses, having a significant impact on battery life. It has fuelled the need for development and exploration of new technologies that provide low standby power, particularly for static random access memories (SRAMs). As SRAM occupies the major area of a die in system-on-chip (SoC) products and consumes significant amount of power. Hence, active and leakage-power-efficient SRAM designs need to be explored for longer operation of battery operated systems. There are two primary areas having strong potential of active and leakage power reduction [61]: (a) lowering the operating voltage which has quadratic dependency with active power ($P_{active} = \alpha * C * V_{DD}^2 * f$) and linear relationship with leakage power, and (b) reduction in charging or discharging the capacitance of word and bit lines. It has been reported that up to 70% of the total active power is dissipated in charging or discharging of bitlines during read and write operations [109].

Lowering the operating voltage or a sub-threshold operation for active and leakage power saving notably degrades the standard 6T SRAM bitcell's read stability and write-ability noise margins for nano-meter or sub-90 nm technologies. These margins are further degraded by process-variation (or device mismatch) due to shrinking the device feature sizes, to achieve high density and high performance objectives of on-chip caches. Therefore, achieving sufficient bitcell noise margins in the nano-meter regime with optimized device sizes, choosing the right bitcell topologies and design methodologies are important for SRAM bitcell designer. In sub-threshold SRAMs, it becomes harder to ease the difficulty of fabrication process and susceptibility to process variations because of minimum feature sized devices are employed.

J. Singh et al., *Robust SRAM Designs and Analysis*, DOI 10.1007/978-1-4614-0818-5_3,   57
© Springer Science+Business Media New York 2013

Rather than simply down scaling the operating voltage, several Dynamic Voltage Scaling (DVS) techniques were proposed to reduce the active and leakage power of both idle and active memory blocks. The DVS can be achieved by either lowering V_{DD} [11,57,61], or raising the ground, V_{SS} [2], or both [31]. Many bitcell topologies and design methodologies that employ DVS [25, 84, 85] have been proposed in recent past to achieve low energy consumption (by operating at or near sub-threshold region) and high density, while, ensuring sufficient noise margins and susceptibility to process variations.

Some researchers believe that it is necessary to move from standard six-transistor (6T) SRAM bitcell to eight-transistor (8T) or ten-transistor (10T) register-file (separate read and write ports) type of bitcell structures. This shift in SRAM bitcell design is to cope with poor noise margins or the process variability problem when chips to operate at low voltages. Register-file type SRAM bitcell topologies (8T or 10T) de-couple the read and write operations of a standard 6T SRAM bitcell by creating an isolated read-port or read buffer (comprised of 2T or 4T). Register-file type of SRAM bitcells have recently attracting attention [18,23–25,73,103,108,111] because of improved noise margins and less susceptibility to process variation. Some of these bitcell topologies also employ DVS techniques to achieve sub-threshold operation and to minimize the dynamic and leakage power for energy constraint applications, such as medical implants, sensor nodes and hand held devices. However, DVS needs routing and creation of dynamic V_{DD} and V_{SS} using higher metal layers, and DC to DC converters. This use of higher metal layers increase metal density and wire delay or capacitive coupling between adjacent word and bit lines. Thus, increased number of transistors per bitcell and employing DVS techniques may not be advisable for a good SRAM bitcell candidate to achieve high density on-chip caches.

Reduction in charging or discharging capacitance of a word or bit line can be achieved either by having fewer word or bit lines for accessing a bitcell, or by the partial activation of multi-divided word or bit lines. It has two advantages, apart from saving of active and leakage power: (a) use of fewer word and bit lines reduces the silicon overhead and (b) a divided word or bit line configuration will reduce the wire delay. In an 8T or 10T bitcell, read stability concern is eliminated (SNM-free read) because of separate read and write ports (1R/1W), and they are controlled by separate read and write wordlines. However, during a write event, bitcell disturbance can also occur in an unselected column when a wordline is activated and bitlines are held high, *called partial write disturbance* (PWD) [25]. This PWD situation is identical to that existing in 6T bitcell during read event. An 8T or 10T bitcell can be easily optimized to withstand PWD problem because of separate read and write ports. But the benefits of SNM-free read stability and moving to smaller geometries (technology scaling) would be lost due to up sizing of the devices. In order to achieve the full benefits of SNM-free read stability and technology scaling, PWD must be eliminated without sacrificing the 8T or 10T advantages and over-sizing the devices. In order to circumvent PWD, modification in array organization is needed such that the column select functionality within the SRAM array can be disallowed.

For high density cache memories minimum number of transistors, bitlines, wordlines per bitcell are usually preferred. In this line, single-ended (SE) five-transistor (5T) SRAM bitcell is a good candidate, however, it suffers from sever read and write stability problems. Also in single-ended SE-SRAMs, the use of a single bitline (or data-line) for reading and writing reduces the switching activity factor of a bitline to less than '0.5' as compared to standard 6T, 8T and 10 SRAM bitcells [4], which has strong potential of saving both active and leakage power. The major drawbacks of SE-SRAM bitcells are poor noise margins and difficulty of writing a logic '1' through a NMOS pass-gate device. As a result, optimization of the bitcell size for both read and write operations become non-trivial, since a single pass-gate device is used for both reading and writing operations [4, 44, 113]. Therefore, SE-SRAMs are not widely accepted in high performance cache memories.

3.2 SRAM Bitcell Topologies

In this section, some of the emerging SRAM bitcell topologies will be discussed such as transmission gate based single-ended-6T (TG-6T) [120], register-file type 8T bitcell [108] and single-ended 6T SRAM bitcell [98]. An exhaustive study on SE-6T SRAM bitcell and comparison with TG-6T and 8T bitcells will be presented. These bitcell topologies perfectly fit in sub-threshold and low power applications. A comprehensive study of these designs and their merits and demerits are also highlighted.

3.2.1 Transmission Gate Based Access 6T (TG-6T) SRAM Bitcell

The schematic diagram of a transmission gate (TG) based 6T SRAM bitcell hereafter referred as a 'TG-6T', is shown in Fig. 3.1 [120]. In SE bitcells [4,44,113], accessing and transferring of data is performed through a NMOS pass-gate device. However, the TG-6T employs a transmission gate for accessing and transferring the data during read and write event via single-ended bitline (data line). Employing a TG for reading from and writing into a memory bitcell is dangerous because it conducts perfectly for '0' and '1' in both directions. In other words, a read event can easily flip the bitcell content, because of direct and strong intervention of read-current (shown in dotted) to data storage node Q, as shown in Fig. 3.1. Hence, the use of TG drastically reduces the read stability margin (or read SNM), to even worse than the standard 6T SRAM bitcell, obtained from the butterfly curve, as shown in Fig. 3.3.

In a TG 6T bitcell, all the devices have to sized too large to achieve adequate read SNM, making the bitcell size more than twice the standard 6T bitcell, thus defeating

Fig. 3.1 Schematic diagram
of a transmission gate based
access transistor SRAM
bitcell (TG 6T) [120], with
read-current path (shown in
dotted line)

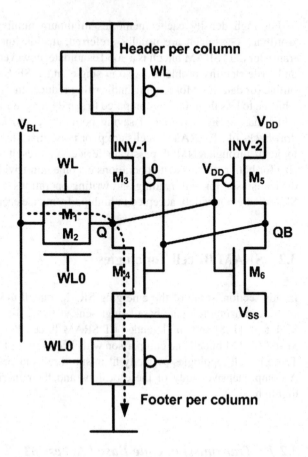

the benefits of technology scaling. The write operation in single ended SRAM
bitcells is difficult because of strongly cross-coupled inverters and a NMOS pass-
gate device is used for writing operation (transferring the data to storage nodes).
Several write assist circuits and the weakening of cross-coupled inverters techniques
have been proposed to overcome this problem [4, 44]. In a TG 6T bitcell, both
read and write operations are assisted by a header and footer per column to achieve
good read SNM and write-ability. However, sharing a header and footer per column
or bitline will affect the stability of the other (non-accessed) bitcells connected to
the same bitline, hence the column select functionality in the TG-6T SRAM array
organization may worsen the PWD problem. Another worth noting drawback is that
this bitcell topology limits the number of bitcells connected to one bitline, BL, to
16. Therefore, applications requiring more than 2 kb of memory, the area overhead
of peripheral circuitry might limit its practicality.

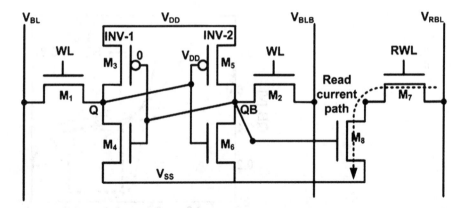

Fig. 3.2 Schematic diagram of an isolated read-port 8T [108] SRAM bitcell, with read-current path (shown in *dotted line*)

3.2.2 Separate Read-Port 8T SRAM Bitcell

Figure 3.2 shows a 2-port (1-Read/1-Write) 8T bitcell [108], with a separate read-port comprising of two transistors M_7 and M_8, and a standard 6T bitcell having separate write access NMOS pass-gate devices (M_1 and M_2). In this design, read operation is performed with the activation of single-ended read bitline (V_{RBL}) and read wordline (RWL). However, write operation in this design is differential in nature and performed with the help of separate write wordline (WL) and bitlines (V_{BL} and V_{BLB}). Write operation of this design is performed in similar to standard 6T SRAM bitcell. A 1R/1W port or separate read and write ports mechanism offers a static-noise-margin-free read (SNM free read) operation, because it isolates the read-current path (shown in dotted) from the data storage nodes (Q and QB). Hence, it eliminates the conflicting read and write requirements of sizing of read and write pass-gate devices which exist in standard 6T and TG-6T bitcells. In other words, it makes bitcell size optimization more easier. Thus, in a 8T bitcell the optimization of device size can be done separately for target margins to achieve a delicate balance between read-stability and write-ability margins. This feature widens the bitcell optimization space to a large extent by ensuring sufficient bitcell margins.

Employing separate read-ports provide more than two times better read SNM that cannot be achieved in standard 6T and TG 6T bitcells, even with dynamic voltage scaling (DVS) techniques, while maintaining a strong write-ability of logic '1', as shown in Fig. 3.3. One can observed that the read SNM of a SE 6T and 8T SRAM bitcells is about 2.5× better than the standard 6T and TG 6T SRAM bitcells. In subthreshold SRAM bitcell design, there are many challenges namely poor read SNM, weak write-ability and increased bitline leakage current in the presence of process variations. These challenges can be overcome with read and write assist circuitries. For instance, the read-SNM is eliminated by a separate read-port [23] mechanism, while the write-ability is enhanced by a virtual supply node VV_{DD}.

Fig. 3.3 The voltage transfer characteristics and read SNM comparison for standard 6T, TG 6T, 8T and SE 6T SRAM bitcells

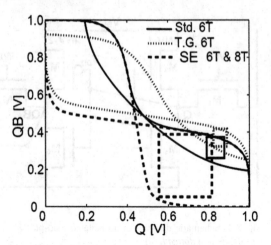

However, the bitline leakage current problem is solved by using a read-buffer foot which is shared among the bitcells of a word. The feet of all the read-buffers in a row are connected to the read-buffer footer (a peripheral driver). This read-buffer footer pulls feet of all the read-buffer nodes low only in the case of accessed row, otherwise pulls at high [108]. A non-interleaved layout is employed where the accessed bitcells are kept together in order to share a read-buffer footer and VV_{DD} among the accessed bitcells.

3.3 Single-Ended 6T SRAM (SE-SRAM) Bitcell

Figure 3.4 shows the single-ended separate read-port 6T SRAM bitcell design (read-current path is shown with dotted line) [98]. This design consists of a cross-coupled inverter pair (INV-1 and INV-2) connected in a positive feedback. The inverter pair is connected to a read or write bitline (V_{BL}) by an access transistor (M_1) which is controlled by a write wordline (WWL). A separate read-port used for reading the content of the bitcell comprises of transistor $M1_R$ and $M2_R$. The read or write bitline (V_{BL}) is connected to ground if the node QB and RWL are high, which correspond to '0' bit is stored in the bitcell. Similarly, if node QB is low and RWL is high, V_{BL} is approximately kept high, which correspond to '1' bit is stored in the bitcell. Transistor $M2_R$ in read-port is controlled by the read wordline (RWL) to read the bitcell content, and it is shared per word. Similarly, a write ($M1_W$) assist transistor is also shared per word.

The shaded transistors shown in Fig. 3.4 ($M2_R$ and $M1_W$) are read and write assist transistors, respectively, for a memory word and shared among all the bitcells in that word. The memory word can be 8, 16, or 32 bit etc. The single-ended and separate read-current path mechanisms in the SE 6T bitcell design offers both the advantages,

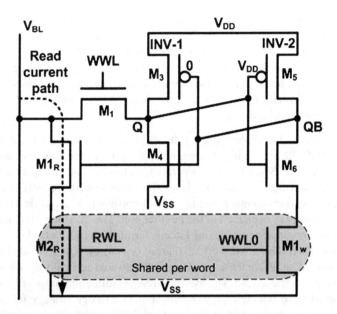

Fig. 3.4 Schematic diagram of a single ended (SE) 6T SRAM with shared read and write assist transistors per word and read-current path (shown in *dotted line*)

power saving and subthreshold operation. Power saving in the this design comes from two places: operating at lower voltages, and reduction in the bitline activity factor. However, the challenges of subthreshold operation such as read-SNM, write-ability and bitline leakage in the presence of process variations are handled carefully without increasing the area overhead. The *distinct features of the SE 6T SRAM bitcell as compared to the previously proposed TG 6T and 8T bitcells* [108, 120] are as follows:

- In contrast to 8T SRAM bitcell, SE 6T SRAM bitcell has a single bitline (V_{BL}) is used for both read and write operations. As a result, reduced activity factor and area overhead. The write operation in SE 6T design is assisted by a write ($M1_W$) assist transistor which is shared among all the bitcells of a word; *remember not per column, as header and footer are shared in the TG 6T SRAM bitcell.*
- The separate read-current path comprises of a single transistor ($M1_R$), while another read assist ($M2_R$) transistor is shared by all the bitcells in that word; *not per column.* This arrangement provides a SNM-free read operation because of isolated read-current path from the data storage nodes (Q and QB) and uses minimum number of transistors.
- Swapping the control (gate) terminals of read-current path transistors with respect to an 8T bitcell read-port to minimize the leakage current from the bitline, BL, by the un-accessed bitcells connected to the same BL. This also helps in reducing the loading effect on the data storage nodes due to forward gate tunnelling current, which is quite significant in nano-regime.

- A non-interleaved array organization is employed to facilitate the sharing of read and write assist transistors. It also provides SNM-free read operation and strong write-ability margin simultaneously by eliminating the PWD.
- There are no separate dynamic biases, DC to DC converter or header or footer to add extra overheads. Also, read or write assist transistors are shared per word in order to keep check on array area overhead and disallow the column select functionality within the array.

In SE 6T SRAM bitcell design, the primary concern to operate it in subthreshold region is a tradeoffs among bitcell area, noise margin, read access time and bitline leakage current. The SNM-free read operation comes because of separate read and write current paths, as a result, it relaxes the noise margin tradeoff. Hence, the remaining tradeoffs are tightly related to size of transistors $M2_R$, while $M1_W$ has no specific constraints as explained in the next section. The $M2_R$ and $M1_W$ transistors are shared by a group of neighbouring bitcells forming a word, and to keep check on area penalty. Therefore, a word-oriented array organization with divided wordline has been adopted, in which these transistors are activated vertically by sub-wordline drivers to read or write a word. The use of divided wordline and sub-wordline drivers in the adopted word-oriented array organization is a design strategy for achieving SNM-free read operation and strong write-ability margin simultaneously, while eliminating the PWD problem. However, multi-divide word and bitline techniques are commonly used to reduce the charging and discharging capacitance, or in other words to minimize the wire delay for improving the array performance [47, 119]. The noise margins, bitline leakage, array organization and read access time tradeoffs are explored in detail in subsequent sections.

3.3.1 Array Organization of SE 6T SRAM Bitcell

In word-oriented SRAM array design with the SE 6T SRAM bitcell, a word has non-interleaved more than 1-bit per word. If n be the number of bitcells in a word-oriented SRAM block which should contain more than 1-bit per word, that is, $n \geq 2$. For instance, a word is organized with SE 6T bitcell has $n = 32$ bitcells, as shown in Fig. 3.5. During read or write operation these n bits of a word are simultaneously accessed, therefore, one could share the read and write assist transistors ($M2_R$ and $M1_W$) of a bitcell, as shown in Fig. 3.5. As a result, only one read and write assist transistor per word are needed. Consequently, each bitcell in a word has six transistors with two additional shared transistors per word. Also reading or writing of a word (bitcells) is not affected, when another word is accessed for writing or reading, because a word shares read or write assist transistor by row, not by column.

Proper sizing of read and write assist transistors is crucial because functioning and performance as a whole of a memory array depend upon these transistors. If the sizes are overestimated, then there is a wastage of valuable silicon area, increase in leakage, and switching power dissipation. Also the underestimated sizes would

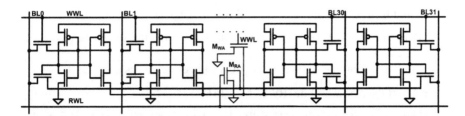

Fig. 3.5 The schematic diagram showing a 32-bit word-oriented array organization of a SE 6T SRAM bitcell with shared read and write assist transistors (M_{RA} and M_{WA}) placed at the center of the array

slow down the read and write operations significantly due to the increased resistance to ground. The purpose of both read and write assists transistors is fundamentally different because read assist transistor has to provide a low resistive path to read current only during read operation. On the other hand, write assist transistor has to provide high resistance path for a successful write operation to weaken the cross-coupled inverter pair. As read and write operations requirements are conflicting in nature, there is a need to analyze the sizing issues separately for read and write assist transistors. The sizing issues of read and write transistors and their impact on the SRAM array size and performance are exhaustively studied in Sect. 3.5.

3.3.2 Read Operation

In SE 6T SRAM bitcell, read operation is performed via single-ended bitline (data-line), that is, information is sensed directly from the bitline. However, in standard 6T SRAM bitcell read operation is performed via sensing the differential voltage exist between bitline and its complement. Without loss of generality, it is assumed that the internal data storage nodes QB and Q are at '0' and '1', respectively, as shown in Fig. 3.4. To read the SE 6T bitcell contents, the following sequence of steps are performed:

- Prior to read operation bitline (BL) is precharged to the supply voltage (V_{DD}).
- The read wordline (RWL) is asserted to high, write wordline (WWL) is kept low and its complement (WWL0) is also asserted to high.
- Rise of RWL from '0' to '1', results activation of read-port.

  - If one of the bitcell side (node QB) stores the logical '1'; the bitline (BL) is discharged through read-port comprises of transistor $M1_R$ and $M2_R$.
  - Similarly, if the bitcell side (node QB) stores the logical '0'; the bitline (BL) will not discharged through read-port because transistor $M1_R$ will be in cut-off state.

- Depending upon whether the bitline BL is discharged or remained at V_{DD}:

  – If bitline BL is kept high (or remained at V_{DD}), the bitcell is read as a logical '1'.
  – Similarly, if bitline BL is discharged, the bitcell is read as a logical '0'.

- De-assert the read wordline (RWL) back to 0.

In both cases, either reading '1' or '0', data storage node (Q) is isolated from the read current path. It avoids the rise in potential of node Q if it stores '0', in contrast to standard 6T SRAM bitcell in which node holds '0' potential rises and limited to the trip-point of an inverter. The separate read-port reduces the capacitive coupling noise induced from BL, and hence, significantly enhances the data stability during read and hold states.

3.3.3   Write Operation

It is a fact that the write operation in single-ended SRAM bitcells is difficult as compared to its counterpart standard 6T bitcell. In standard 6T SRAM bitcell, write operation is aided by both bitline and its complement. However, in SE 6T SRAM bitcell write operation is performed via a single bitline and it is not aided by other bitline, as a result, it is hard to flip the state of the cross-coupled inverters through single bitline. To overcome this problem, a write assist transistor $M1_W$ is used, which is controlled by WWL0. The usage of $M1_W$ is to weaken the cross coupling in the SE 6T bitcell inverters during write access time. The write operation or flipping the bitcell contents when initially assuming that the internal data storage nodes QB and Q are at '0' and '1', respectively, as shown in Fig. 3.4, consists of the following sequence of steps:

- Initially, write wordline (WWL) $= 0$.
- Precharge the bitline (BL) to the supply voltage (V_{DD}).
- After prechrage, bitline (BL) is disconnected from the supply voltage (V_{DD}).
- Write wordline (WWL) is activated to high and its complement (WWL0) is low which makes the cross coupled inverters weak (data enters the bitcell during this step).
- Place the data value on the BL which is connected to the data storage node Q via M_1 and depending upon the input data:

  – If input data is '0' and node Q at '1', bitline BL is driven to ground by a write driver, till the node Q pulls below the trip point of INV-2. Once it reached to trip point of an inverter write operation is done.
  – If input data is '1' and node Q at '0', bitline BL is kept high by a write driver, till the node Q pulls above the trip point of INV-2. Once it reached to trip point of an inverter write operation is done.

- De-assert the write wordline (WWL) back to low and WWL0 to high.

3.4   Read Stability and Write Ability Margins

The hold SNM, read SNM and write trip-point-voltage are commonly used metrics to determine the hold stability, read stability and write-ability of a SRAM bitcell, respectively. The competency of operating at lower voltage and process variation tolerance can also be best judged by these metrics. In order to understand which SRAM bitcell design has better stability and process variation tolerance different SRAM (SE 6T, standard 6T and 8T) designs are considered and their stability metrics are compared. The predictive technology model (PTM) of 65 nm CMOS technology node [88] was used in the SPICE simulations. Parasitics of a small $16 \times 16 \times 32$ bit SRAM array was extracted and incorporated in the SPICE net-list for performing transient and AC simulations. The effect of process variation is considered with 3-σ statistical variations in threshold voltage V_{TH} of all the transistor in each bitcell.

3.4.1   Read Stability Margin (SNM)

Read SNM measured from the voltage transfer characteristics of SE 6T, standard 6T and 8T SRAM bitcells are shown in Fig. 3.6, to present a comparative perspective. The SE 6T and 8T bitcells have SNM of 0.302 V, while the standard 6T bitcell's SNM is 0.152 V at a supply voltage of $V_{DD} = 1.0$ V and $\beta = 2$ for identical (iso-performance) read-current. The read current in the standard 6T has direct interference to the data storage nodes (Q and QB), while the SE 6T and 8T have separate read-current port which eliminate an interference to the data storage nodes.

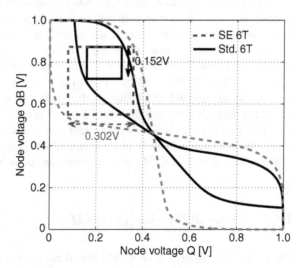

Fig. 3.6 Read static noise margin (SNM) comparison of a standard 6T and SE 6T bitcells during read operation at $V_{DD} = 1$ V

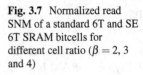

Fig. 3.7 Normalized read SNM of a standard 6T and SE 6T SRAM bitcells for different cell ratio ($\beta = 2$, 3 and 4)

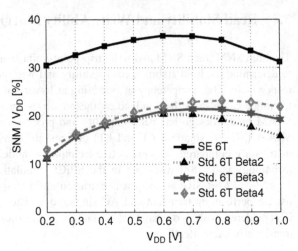

As a result, higher read SNM is obtained for SE 6T and 8T SRAM bitcells as compared to standard 6T SRAM bitcell. The SE 6T and 8T bitcells yield SNM-free read events. In other words, their read SNMs are equivalent to their hold SNMs.

The read SNM normalized to supply voltage V_{DD} for different cell ratio ($\beta = 2$, 3 and 4) of a standard 6T, however, minimum sized devices are used in SE 6T and 8T SRAM bitcells is shown in Fig. 3.7. It shows that the V_{DD} sensitivity of SNM in the SE 6T and 8T bitcells is smaller than that of the standard 6T bitcell. SNMs of the SE 6T and 8T bitcells at a supply voltage of 0.3 V are equal to those of the standard 6T bitcell at 0.5 V and $\beta = 4$. Hence, the SE 6T bitcell with minimum sized devices has better read SNM as compared to its counter part standard 6T bitcell even at lower voltages.

Under process variation, the read SNM of the standard 6T SRAM bitcell degrades drastically due to minimum feature sized devices, conflicting read and write requirements, read-current interference to the data storage nodes and capacitive coupling noise induced from the bitlines. Figures 3.8 and 3.9 show the variation in VTCs and worst case read SNM of the standard 6T, SE 6T and 8T bitcells, respectively. The SE 6T and 8T bitcells provide 2.65 × higher worst-case read SNM as compared to the standard 6T SRAM bitcell under same process variation. Higher worst case read SNM in the SE 6T and 8T bitcell is mainly due to separate read and write (1R/1W) ports, or isolated read-current path which facilitate easy optimization of bitcells for target margins. Hence, *the SE 6T SRAM bitcell preserves both the benefits of scaling as well as SNM-free read event, even in worst case process variations.*

3.4.2   Write-Ability Margin (WAM)

The write-ability margin of an SRAM bitcell can be best characterized by the write trip-point-voltage which is defined as the maximum amount of voltage on bitline

Fig. 3.8 Monte Carlo
simulation of voltage transfer
characteristics (VTCs) shown
with worst case read SNM
during read operation for a
standard 6T SRAM bitcell

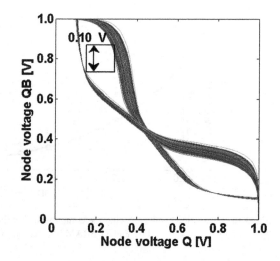

Fig. 3.9 Monte Carlo
simulation of VTCs shown
with worst case read SNM
during read operation for the
SE 6T and 8T SRAM bitcells

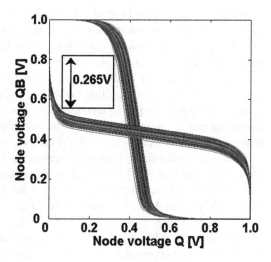

needed to flip the bitcell content [36]. Due to the asymmetric nature of the SE 6T
SRAM bitcell, both the states for a writing '1' and '0' are analyzed. In order to
write a '1' ($Q = 1$ and $QB = 0$) to a bitcell, when initially storing a '0' ($Q = 0$ and
$QB = 1$). The low internal node Q of the bitcell must be pulled up by the precharged
bitline (BL) above the trip-voltage of INV-2. Pulling up or down the data storage
node (Q and QB) in the cross-coupled inverter pair with positive feedback is hard,
owing to regenerative action of the inverters. To circumvent this problem, write
assist techniques were employed in the previously proposed bitcells, such as in TG
6T bitcell, header and footers were used, while in 8T bitcell virtual VV_{DD} supply
was employed. However, in the SE 6T *a minimum feature sized NMOS transistor*
$M1_W$ is employed to slow down the regenerative action of the cross-coupled inverter
pair during the write event.

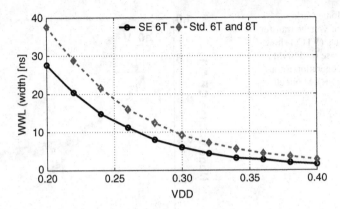

Fig. 3.10 Simulation results showing the comparison of write pulse width for a successful write operation in SE 6T, standard 6T and 8T SRAM bitcells for different V_{DD}

Writing '0' (Q = 0 and QB = 1) to a bitcell, when initially storing a '1' (Q = 1 and QB = 0), high internal node Q of the bitcell has to discharge the bitline (BL) via write driver well below the trip-voltage of the INV-2. To guarantee that a correct write operation will occur, it is important to note that the node Q must be pulled up (or down) above (or below) the trip-voltage of INV-2 within the write access time i.e. when the write wordline (WWL) is high, otherwise a write failure will occur. As shown in Chap. 2, dynamic stability Section, the wordline pulse width has significant role in ensuring a correct write operation. The size of the pulse width is analysed for a write operation in SRAM bitcells, as shown in Fig. 3.10.

For a successful write operation at different V_{DD}, different sizes of write wordline pulse is needed. Figure 3.10 illustrate the minimum WL/WWL pulse width (access time) needed for a successful write operation for different (SE 6T, standard 6T and 8T) SRAM bitcells. It can be observed from Fig. 3.10 that the write operation of the SE 6T bitcell at $V_{DD} = 0.30$ V is 36% faster than the standard 6T and 8T bitcells. In other words, to achieve iso-write-performance at $V_{DD} = 0.30$ V, standard 6T and 8T bitcells have to operate at $V_{DD} = 0.35$ V. Improvement in write performance of a SE 6T bitcell is due to breaking of regenerative action during write cycle with the help of read assist transistor.

The write trip-point-voltage of the SE 6T and 8T bitcells are shown in Figs. 3.11 and 3.12, respectively. The simulation results demonstrate the write-ability of the SE 6T and 8T bitcells at $V_{DD} = 0.3$ V for writing $0 \rightarrow 1$. Writing $0 \rightarrow 1$, is shown here because in the SE 6T writing $1 \rightarrow 0$ is easier and quicker. In standard 6T and 8T SRAM bitcell write operation is symmetric, therefore, writing $0 \rightarrow 1$ or $1 \rightarrow 0$ has no difference. The write trip-point voltage of the SE 6T is 13% (148 mV) higher than 8T bitcell where a virtual VV_{DD} scheme (write assist) has been used. However, it has an advantage that an erroneous write will not take place easily compared to 8T [108], due to bitline noise.

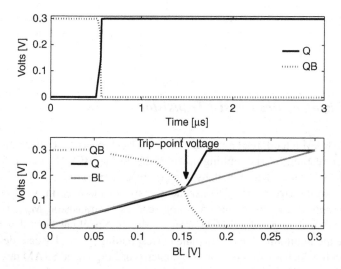

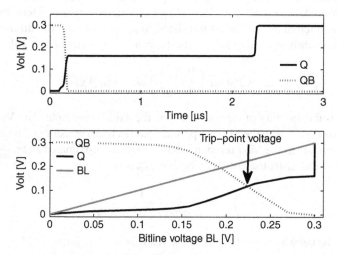

Fig. 3.11 Transient SPICE simulation of write-trip voltage of the SE 6T bitcell at $V_{DD} = 0.3$ V for writing $0 \rightarrow 1$

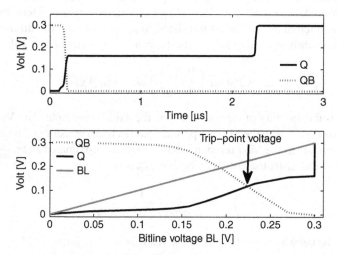

Fig. 3.12 Transient SPICE simulation of write trip-voltage of the standard 6T and 8T [108] bitcell at $V_{DD} = 0.3$ V for writing $0 \rightarrow 1$

3.5 Sizing of Read and Write Assist Transistors in SE 6T Bitcell

Using SE 6T SRAM bitcell, realization of an area efficient SRAM array for a target performance is solely depends on proper sizing of read and write assist transistors. Since overestimated read and write assist transistors will increase the

area overhead, leakage and switching power dissipation. While underestimated size
of read and write assist transistors may significantly hamper the array performance
due to increased ground resistance.

3.5.1   Sizing of Read Assist Transistor

In SE 6T SRAM bitcell, read assist transistor ($M1_RA$) forms the critical path essen-
tially when reading '0' (i.e. pulling down of the bitline, BL). Hence, performance
of SE 6T SRAM is determined by the '0' read access time, which is mainly
dependent on the size of M_{RA}. Consequently, size of the M_{RA} in word-organized
SRAM design when a word has common read assist transistor, M_{RA}, is critical
for proper functioning of an SRAM array. A simple model has been developed to
determine the minimum size of M_{RA} and corresponding '0' read access delay for a
single bitcell, which is extended for the adopted word-organized SRAM design. The
developed model is inspired by well established power gating techniques in which
sleep transistor is used to gate the power supply [51]. In the literature [51, 52], it
was shown that the sleep transistor can be approximated as a linear resistor to create
a virtual ground, because $V_{DS} < (V_{GS} - V_{TH})$ of a sleep transistor. Here, this sleep
transistor is referred as read assist transistor, M_{RA}. The amount of current flowing
through the linearly-operating M_{RA} transistor can be approximated as [6]:

$$I_{RA} \approx \mu_n C_{OX} \left(\frac{W}{L} \right)_{RA} (V_{DD} - V_{TH}) V_{RA} \tag{3.1}$$

where, μ_n is the mobility of electrons, C_{OX} is the oxide capacitance and V_{TH} is the
threshold voltage. Since, the M_{RA} is approximated as linear resistor and operating in
a linear region, then the M_{RA} resistance R_{RA}, can be approximated as $\approx \frac{V_{RA}}{I_{RA}}$. Thus,
the size of a read assist transistor can be expressed as:

$$\left(\frac{W}{L} \right)_{RA} = \frac{1}{R_{RA} \mu_n C_{OX} (V_{DD} - V_{TH})} \tag{3.2}$$

If R_{RA} is known, then the size of a read assist transistor $(W/L)_{RA}$ can be
determined by using the above expression Eq. 3.2. The M_{RA} affects only high to
low transition or reading '0' to discharge the precharged bitline. Since, bitline
capacitance C_{BL} is discharging, and neglecting the node V_{RA} parasitic capacitance,
any charge flowing out of the source of $M1_R$ will flow through the read assist resistor
R_{RA} of M_{RA}. This phenomenon is modeled as a *R-C circuit*, which comprises of
series resistor R_{RA} and bit line capacitance C_{BL} charged at voltage V_{DD}. Thus, the
relationship among these parameters can be expressed as follows:

$$V_{RA} = V_{DD} \times exp \left(\frac{-t}{\tau} \right) \tag{3.3}$$

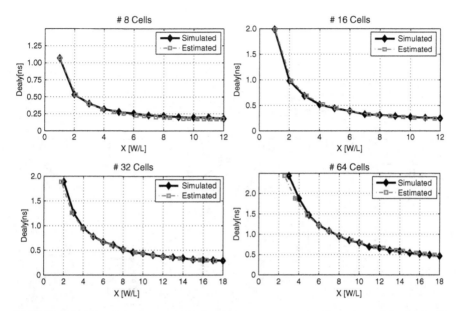

Fig. 3.13 Estimation of read access delay for different read assist transistor sizes (W/L)

Where, τ is the time constant. If the read sensing circuitry detects the transition high to low i.e. read '0', only when the bitline is discharged to 36.8% of the V_{DD} after a certain amount of delay from the assertion of read control signal, which is defined as a read access delay. Under this condition the read access delay τ_d is equal to time constant (τ):

$$\tau_d = R_{RA}C_{BL} \tag{3.4}$$

In the word-organized SRAM array, as shown in Fig. 3.5, let, the word be n-bit wide, that is there are n-bitcells in each word and all are having individual M_{RA}. These individual M_{RA} of n-bitcells in a word are replaced by an equivalent M_{RA} to reduce the transistor count and silicon area overhead. The size of M_{RA} in worst case pattern (i.e. when all the n-cells having '0' at node Q) determines the read access delay or operating frequency of an SRAM array. As we have approximated the M_{RA} of a bitcell as a linear resistor, then all the n-bitcells M_{RA} will form a parallel combination of n-linear resistors in worst case pattern. In this case, the M_{RA} resistance will be equivalent to M_{RA}/n. Similarly, n-precharged bitlines capacitance (neglecting the node capacitance) will be replaced by an equivalent capacitance nC_{BL} because of parallel combination they form. Once, an equivalent resistance, capacitance and target read access delay are known then from Eqs. 3.2 to 3.4, size of the M_{RA} for any word size can be determined easily.

The SPICE simulation and mathematical model based estimated results for a read assist transistor size (W/L) and read access delay for different word sizes ($n = 8, 16, 32$ and 64) of a word-organized SRAM design are shown in Fig. 3.13. It can observed that the developed mathematical model achieves up to 95% accuracy

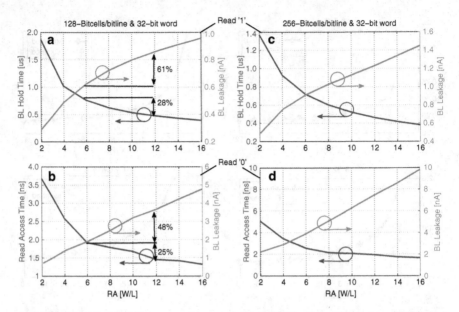

Fig. 3.14 (**a**) Bitline leakage current from unaccessed 127-bitcell per bitline for a 32-bit word and read '1' bitline hold time for different sizes of shared read assist transistor, M_{RA}. (**b**) Bitline leakage current from unaccessed 127-bitcell per bitline for a 32-bit word and read '0' access time for different sizes of shared read assist transistor, M_{RA}. (**c**) Bitline leakage current from unaccessed 255-bitcell per bitline for a 32-bit word and read '1' bitline hold time for different sizes of shared read assist transistor, M_{RA}. (**d**) Bitline leakage current from unaccessed 255-bitcell per bitline for a 32-bit word and read '0' access time for different sizes of shared read assist transistor, M_{RA}

in estimation of read access delay for different word sizes. It is clear from the above discussion and derived mathematical model that the size of M_{RA} has a strong relationship with performance (i.e. '0' read access time). At the same time bitline leakage current will increase due to an additional read-port and it is directly related to the size of the read assist transistor in a read-port. A large bitline leakage current from the un-accessed bitcells may result an erroneous read operation.

In order to establish a tradeoff among bitline leakage current, read access time (performance) and the size of the read assist transistor, M_{RA}, simulations of 32-bit and 64-bit words size with 128 and 256 bitcells connected per bitline are performed. Figure 3.14a, c show the bitline leakage current and discharge (bitline hold) time for the worst case read '1' from a 32-bit word, when all the un-accessed bitcells of a 32-bit word hold '0' with 127-bitcells per bitline and 255-bitcells per bitline, respectively. The bitline hold time is measured, when all the bitlines (BLs) of a word are precharged to V_{DD} and RWL is asserted high to sense the logic '1' from the bitlines, before BLs discharged to $0.5V_{DD}$ via M_{RA}, due to leakage current from the unaccessed bitcells. This leakage current also include the leakage from the read-assist line shared by all the bitcells in a word.

If the RWL pulse width is higher than the bitline hold time, then there will be an erroneous read event. As the size of M_{RA} increases, the bitline leakage current

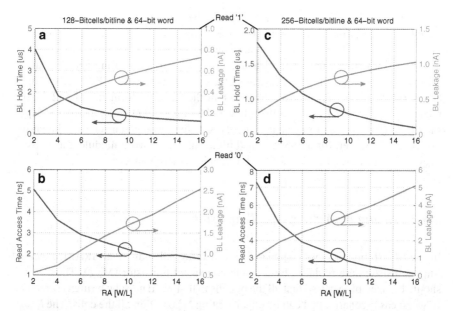

Fig. 3.15 (**a**) Bitline leakage current from unaccessed 127-bitcell per bitline for a 64-bit word and read '1' bitline hold time for different sizes of shared read assist transistor, M_{RA}. (**b**) Bitline leakage current from unaccessed 127-bitcell per bitline for a 64-bit word and read '0' access time for different sizes of shared read assist transistor, M_{RA}. (**c**) Bitline leakage current from unaccessed 255-bitcell per bitline for a 64-bit word and read '1' bitline hold time for different sizes of shared read assist transistor, M_{RA}. (**d**) Bitline leakage current from unaccessed 255-bitcell per bitline for a 64-bit word and read '0' access time for different sizes of shared read assist transistor, M_{RA}

from the un-accessed bitcells increase almost linearly, while the bitline hold time decreases quite slowly. For instance, there is a 61% increase in bitline leakage current and 28% decrease in bitline hold time, if the size of a M_{RA} is increased from $W/L = 6$ to $W/L = 12$, as shown in Fig. 3.14a, c. A similar trend is also observed if the size of a word (i.e. a 64-bit word) is doubled. However, it reduces the bitline leakage and increases the bitline hold time as compared to a 32-bit word, as shown in Fig. 3.15a, c.

In Fig. 3.14b, the optimal point with respect to performance and bitline leakage current is corresponds to $W/L = 6$, where '0' read access time is 2 ns, which determines the operating frequency and bitline leakage current is 2 nA, due to the unaccessed 127-bitcell/bitline. If width of M_{RA} for a word is increased, the rate of increase of bitline leakage is almost double the rate of decrease of '0' read access time. For instance, increasing the width of M_{RA} ($W/L = 12$) will increase the bitline leakage current by 48%, while, the '0' read access time will reduced by only 25% as compared to $W/L = 6$. Hence, increasing the width of M_{RA} beyond this point may lead to a large bitline leakage current. As the bitcells/bitline is doubled, as shown in Fig. 3.14d, a similar trend is observed, while bitline leakage current from the un-accessed bitcells is almost doubled and a slight drop in '0' read access time due to increased bitcells/bitline. However, to achieve the equal performance with doubled bitcells/bitline (256-bitcells/bitline), the size of M_{RA} ($W/L = 8$) requires to increase to discharge the bitline with almost the same rate.

Figure 3.15 shows the bitline leakage current, read access time and discharge (bitline hold) time for the worst case read, for a 64-bit word with 128-bitcells/bitline and 256-bitcells/bitline. In order to maintain the same performance (i.e. 2 ns '0' read access time) for a 128-bitcells/bitline and 64-bit word module, the size of M_{RA} should be ($W/L = 12$) to discharge the 2 nA bitline leakage current, as shown in Fig. 3.15b. However, for a 255-bitcells/bitline and 64-bit word module, the size of M_{RA} should be more than ($W/L = 16$), for an iso-performance.

3.5.2   Sizing of Write Assist Transistor

In the SE 6T bitcell word-organized SRAM array, all individual SRAM bitcell's M_{WA} transistors are replaced by a single equivalent transistor (M_{WA}). Thus, M_{WA} should be sized properly so that all the bitcells in that word must be written correctly. In worst case scenario, that can be either writing '1' or '0' in all the cells. The M_{WA} has to weaken the cross coupled inverters by floating the INV-2 of all the bitcells in that word. Weakening of the loop doesn't matter whether it is intend to write '1' or '0' in all or fewer bitcells in that word. The weakening of the loop of a single bitcell or all the bitcells in a word is equivalent because V_{DS} of M_{WA} is always higher than the '0', when V_{GS} of M_{WA} is zero. Thus, a minimum sized transistor would be well suited for this purpose. Also after the write access time M_{WA} has to provide a ground to node V_{WA} of all the bitcells. For providing a ground to node V_{WA}, M_{WA} has to provide only the leakage current path to all the bitcells whether they are having '0' or '1' at node Q. Since, the transistor M_6 (when node Q at '0') and transistor M_5 (when node Q at '1') are in cutoff mode, therefore, there is only leakage current has to flow through M_{WA}. As M_{WA} has to provide only the leakage current path to all the bitcells of a word which will always less than the dynamic current of a transistor even when all the bitcells are written either '1' or '0' simultaneously. Also, for minimum leakage and data retention it is recommended to use minimum size of transistor. The SPICE simulation for different word size of SRAM reveals that there is no significant improvement in the write-ability of an SRAM array with increasing the size of M_{WA}.

3.5.3   Floor Plan to Eliminate the PWD

In order to circumvent the problem of Partial Write Disturbance (PWD) while retaining the advantages of technology scaling and SNM-free read stability, a practical array organization technique has been presented. Array organization techniques are as important as the bitcell itself to preserve the nanoscale regime

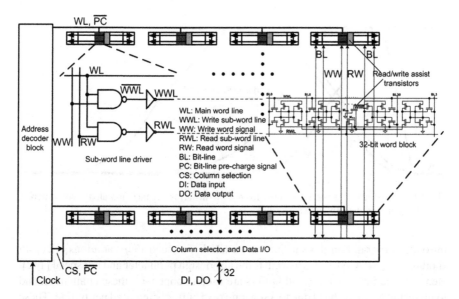

Fig. 3.16 Floor plan of the word-oriented SE 6T SRAM array organization with sub-wordline drivers and a 32-bit word

advantages. In this word-oriented SE 6T bitcell array design, instead of traditional interleaved layout, non-interleaved layout has been adopted in order to avoid the multiplexed column select functionality within the array. This feature facilitates sharing of read and write assist transistors per word to reduce the array silicon overhead, SNM-free read operation and strong write-ability margin simultaneously, while eliminating the PWD problem. These are the key design strategies that motivated for a non-interleaved word-oriented array design. The complete word-oriented floor plan with SE 6T SRAM based bitcell is shown in Fig. 3.16.

Figure 3.16 shows a floor plan of the word-oriented SE 6T SRAM array organization. In this word-oriented SRAM array organization, the problem of PWD which exists in standard SRAM array organization, resulting in poor stability during write access, is being addressed and eliminated. The problem of PWD mainly arises during the write event when specific interleaved bitcells of a word in the accessed row get written through a column multiplexor and bitlines are held high. Since, all these specific bitcells are connected to an activated wordline which turns ON the access (pass-gate) devices of all the bitcells, results loss of stability. In this array organization, the PWD problem is addressed by accessing all the bitcells of a word, kept together (i.e. non-interleaved) by a multi-divided wordline architecture without column multiplexing.

As columns are not multiplexed, a separate set of Sense Amplifiers (SAs) for each column is required, making the area of each set of SAs are more constrained. However, shared SAs used in multiplexed column have more constrained in nanometer regime, because of non-scalability of differential designs. Also an SA design is more challenging especially in subthreshold SRAMs, when considering variability

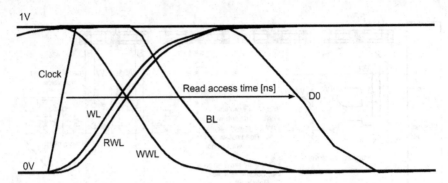

Fig. 3.17 Measured '0' read access time from a SRAM array organization along with different control signals and data output waveforms

introduced by random dopant fluctuation (RDF) induced V_{TH} variations [121]. In subthreshold SRAMs, noise margin is the key design parameter and not speed [112]. Hence, static inverter-type read buffers are used to cope with these challenges, and to offset the area overhead due to a separate read buffer being used per bitline. These read buffers provide the maximum sensing margin for a given supply voltage, due to the full swing in the bitlines.

The use of a non-interleaved word structure and a multi-divided wordline architecture requires a separate set of sub-wordline drivers, adding to the area of the array. However, the main wordline drivers scale with their load, because they have to drive fewer sub-word line drivers and can offset the area overhead incurred by the sub-wordline drivers. The wordline (WL) is assigned to the main wordline, and it is driven by the main wordline driver. The sub-word lines WWL and RWL are assigned to the write sub-wordline and read sub-wordline, respectively, and they are driven by the sub-wordline drivers. These CMOS NAND gates' sub-wordline drivers are controlled by the main wordline WL, and column selection lines, write word, WW and read word, RW, as shown in Fig. 3.16. The SPICE simulated control signals and data output waveforms of a $16 \times 16 \times 32$ bit SRAM module with the SE 6T bitcell are shown in Fig. 3.17. The read access time is measured from 50% of rising edge clock to 50% of output data line. The critical path in the proposed design is '0' read access path which determines the operating frequency. Therefore, we have shown the measured '0' read access time along with other control signals in Fig. 3.17.

3.6 Performance and Power Dissipation

Performance and power dissipation of different SRAM bitcells are presented in this section. Performance is measured in terms of read access time and worst case statistical analysis is done. Dynamic and leakage power dissipation analysis of a small SRAM modules are also done.

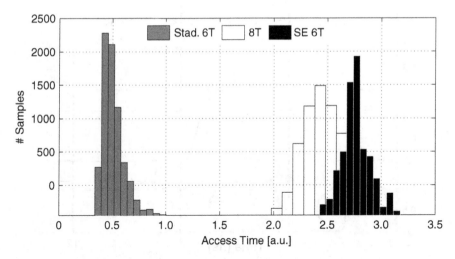

Fig. 3.18 Distribution of read access time due to variation in threshold voltage of SE 6T, standard 6T and 8T SRAM bitcells

3.6.1 Read Access Time Distribution

Figure 3.18 compares the distribution of read access time of the SE 6T, standard 6T and 8T bitcells. The read access time distribution was obtained by the Monte Carlo simulations. Each bitcell was simulated under $\pm 3\sigma$ random variations in threshold voltage of each transistor. For the SE 6T and 8T SRAM read access time was calculated when rising edge of the clock followed by the read wordline (RWL) rises to $0.5 \times V_{DD}$, where, $V_{DD} = 1$ V, to a time when the output of the sense amplifier (read buffer) is reached to $0.5 \times V_{DD}$, as shown in Fig. 3.17. Similarly, in a standard 6T design, read access time was defined as the time between the rising edge of the clock followed by the WL rises to $0.5 \times V_{DD}$ to a time when the expected differential voltage of bitlines (50) mV is achieved.

In the SE 6T bitcell design, '0' read access time has been measured due to its critical path while read '1' access time does not forms critical path. The read access time of the SE 6T bitcell is 10% higher than that of the 8T bitcell, because position of the read-port control transistor have been swapped with respect to 8T bitcell, reducing the read performance of the SE 6T bitcell. The improvement in read performance can be further enhanced by optimizing the size of read assist transistor in SE 6T bitcell. However, the mean read access time of a standard 6T bitcell is significantly lower due to differential read operation as compared to the SE 6T and 8T bitcells that employ single ended read operation.

The distribution of the read access time of a standard 6T bitcell has log-normal dependency with the variation in threshold voltage. Therefore, less number of bitcells fall in tail of the distribution curve. While the SE 6T and 8T bitcells have normal distribution and there are more number of bitcells in the tail of the

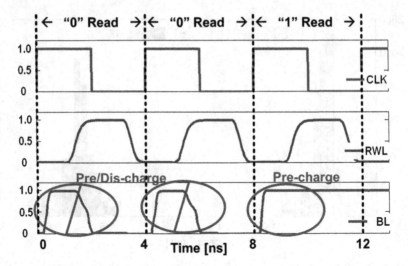

Fig. 3.19 Transient SPICE simulation of bitline (BL) precharging and discharging during read operation of SE 6T and 8T bitcells

distribution curve as compared to the standard 6T bitcell. Furthermore, in worst case scenario to avoid the meta-stability, performance of the SE 6T and 8T bitcell SRAM design is about three time poor than the standard 6T bitcell design. The main reason behind this poor performance is a single ended read operation and increased capacitance due to shared read assist line by all the bitcells in a word.

3.6.2   Power and Leakage Dissipation

Figure 3.19 shows the read operation waveforms of SE 6T and 8T bitcells. Read operation waveforms of both the bitcells are identical, since both employ the single ended isolated read-port mechanism. Hence, bitline power dissipation patterns for read event in both the bitcells are identical. Write operation of standard 6T and 8T bitcells are identical, and hence they have identical write power dissipation pattern from bitlines. However, the SE 6T has a different write mechanism because it employs single ended write via a pass-gate device. Since, most of the active power (up to 70% of the total) is dissipated in charging and discharging of BLs during read or write operation in SRAMs [109]. The pattern of charging and discharging of BLs is a good measure for active power. Figure 3.19 depicts that a certain amount of power is dissipated in 6T by precharging the BL prior to a read operation and discharging only when BL has to changed. However, no power is dissipated by the BL if the upcoming data bit is the same (only for high) as previous one. Similarly, for a write operation, precharging the BL dissipates a certain amount of power for

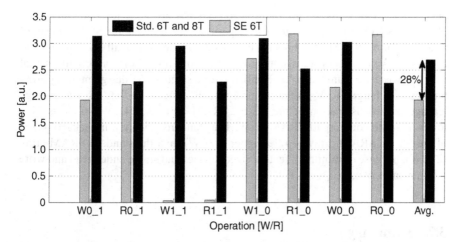

Fig. 3.20 Active power pattern for different read/write operations of SE 6T, standard 6T and 8T SRAM bitcells

each write operation if the upcoming data bit is the same (only for low) as previous one. However, no power is dissipated in BL if the upcoming data bit is the same (only for high) as the previous one.

Figure 3.20 compares active power in the SE 6T, standard 6T and 8T bitcells for different read or write operations. As the SE 6T and 8T bitcells are asymmetric in nature, their active power consumption pattern is also asymmetric. In Fig. 3.20, operation W0_1 stands for writing '1' into the bitcell while its original content was '0'. Similarly, R1_0 stands for reading '0' from the bitcell, while its previous output was '1'. For operation R1_1 the active power of SE 6T and 8T bitcells have been drastically reduced, as compared to standard 6T bitcell, because R1_1 operation is performed without discharging or charging the read bitline of either SE 6T or 8T bitcells. Under such operations, precharged or charged bitline can be used for future read or write operation. Alternatively, in 6T bitcell one bitline has to discharge during these operations. The average active power under different read and write operations in the SE 6T is 28% and 25% lower than the standard 6T and 8T bitcells (Fig. 3.20).

In a $16 \times 16 \times 32$ bit SRAM memory using SE 6T bitcells, reading a word "1110 1110...1110" consumes an average power of only 31% (3.86 mW) as commpared to the standard 6T SRAM memory array because of the reuse of the charged bitline. While, reading a word "0001 0001...0001" consumes 128% (15.94 mW) of the standard 6T SRAM memory. Reading a word with alternating values "1010 1010...1010" uses 68% (8.47 mW) of the standard 6T SRAM memory array power.

The leakage contribution pattern of the SE 6T bitcell is asymmetric due to asymmetric nature of the SE SRAM bitcell. When node $Q = 0$, it leaks more as compared to $Q = 1$ because the read current path transistor $M1_R$ is turned on. The average leakage contribution in the SE 6T bitcell is 37% less than a standard 6T bitcell it is mainly due to minimum sized devices are used for implementation of SE 6T SRAM bitcell. The total leakage in a $16 \times 16 \times 32$ bit SRAM memory array that uses the SE 6T bitcells in standby mode, when all the bitlines are charged to V_{DD}, access transistors (M_1) of a word are cutoff and control signal read and write are clamped at '0'. Similarly, for a standard 6T design memory array bitlines are charged to V_{DD}, and control signals are clamped at '0'. The average leakage power dissipation of the SE 6T SRAM array is 1.4 mW, which is 21% lower than the counterpart SRAM array. The standard deviation in leakage power of the SE 6T SRAM array is 42% higher (32 µW) than the standard SRAM array (23 µW), because of minimum feature sized devices and single ended read and write mechanisms are used.

3.7 Summary

A comprehensive study of SE 6T SRAM and comparison with standard 6T, 8T SRAM bitcells is presented in this chapter. Robustness against the process variation tolerance in SE 6T SRAM bitcell is achieved by isolating the read-current path, or in other words, directly sensing the data from the bitline. Therefore, making both read and write operations independent of each other. The high density in the SE 6T design is obtained by sharing the read and write assist transistors per word, which makes an SRAM array an area efficient. The sharing of these transistors per word instead of per column helps in avoiding the partial write disturbance and un-stability problem to un-accessed bitcells. The dynamic and leakage power with the SE 6T bitcell in the 8 Kb SRAM module are reduced by 28% and 21%, respectively, as compared to standard 6T bitcell SRAM module. The saving in dynamic and leakage power mainly due to use of single ended bitline at the cost of poor performance. The improved read and write-ability, reduced active and leakage power dissipation compared to standard 6T and 8T bitcells makes the new approach attractive for energy constrained applications in the nano-CMOS regime, where process variation is a major design constraint.

Chapter 4
2-Port SRAM Bitcell Design

4.1 Introduction

System on a chip (SoC) products typically contain an increasing number, variety and hierarchy of memories to meet the expected demand of power, area and throughput. Therefore, low power, minimum transistor count and faster access SRAMs are essential for pipelines or parallelism in embedded, multimedia and communication applications which are omnipresent. To accommodate pipelined or parallelism features, simultaneous or parallel read/write access multi-port SRAM bitcells are often employed. The multi-port SRAM bitcell topologies are mainly used to increase the memory bandwidth or simultaneous access of multiple locations within an array. In image or video processors horizontal and vertical data pixels are generally accessed simultaneously to apply different search and process algorithms. The multi-port SRAM bitcells store a single data bit and include two or more access devices in order to provide multi-port access capabilities. These access devices may be connected to the separate read or write wordlines or bitlines, as a result, area overhead increases quadratically with an additional port in multi-ported SRAM bitcells. However exploiting the parallelism phenomenon in multi-port SRAMs in order to improve the bandwidth leads to certain design constraints. Therefore, implementation of multiple access memory bitcells result in large bitcell size, simultaneous read/write disturbance and access conflict, and hence, need to be addressed in the research.

Multi-ported SRAM bitcells are commonly used to implement register file in processors. General purpose processor register file typically wants to handle two reads and one write operation per cycle. Similarly, MIPS must read two sources or write a result on same cycle. A pipelined MIPS must read two sources and write a third result each cycle. However, superscaler MIPS must read and write many sources and result in each cycle.

Aggressive scaling of CMOS technology presents a number of distinct challenges for embedded memory fabrics. For instance, smaller feature sizes imply a greater impact of process and design variability, including random threshold voltage (V_{TH})

variation, originating from the fluctuation in number of dopants and poly-gate edge roughness [75, 104]. The process and design variability leads to a greater loss of parametric yield [10] due to poor SRAM bitcell noise margins and degraded bitcell read-currents, when a large number of devices are integrated into a single die. Therefore, a sufficiently large Static Noise Margin (SNM), Write-Ability Margin (WAM) and read-current (I_{read}) in a bitcell are needed to be maintained carefully to prevent the tremendous loss of parametric yield caused by the technology scaling induced side-effects.

Several SRAM bitcell topologies [18,23] and design methodologies [84,85] have been discussed in the literature for 1-port SRAM bitcells, addressing the nano-regime challenges. However, it is a non-trivial task to simultaneously maintain SNM, WAM, and I_{read} in multi-port SRAM bitcells [101]. In addition, some circuit techniques have been proposed to solve the SNM, WAM, I_{read} and simultaneous access conflict issues in 2-port bitcells [83, 101, 108]. In [83], a priority row decoder circuit and shifted bit-lines access scheme was employed to improve the SNM and eliminate the simultaneous access conflict problem, but this scheme does not fit in independent clocking systems. The isolated (separate) read-port bitcells have recently been the center of attention because of the SNM-free read operation, and improved WAM by providing an additional biasing to the bitcell [101, 108]. A misread (or an erroneous read) problem or large leakage current drawn from the pre-charged bitlines by the un-accessed bitcells limit the number of bitcells per bitline in an array, as a result, it reduces the array efficiency. The erroneous read problem is almost eliminated by the use of read-foot buffer shared among the bitcells per word [108]. However, additional biasing and read-buffer foot lead to an extra silicon overhead and a considerable trade off in floor-planning.

In order to address the above discussed shortcomings of the multi-port SRAM bitcells, in this chapter study of existing 2-port SRAM bitcells is presented and a case study of state-of-the-art 2-port 6T memory bitcell is also presented. In multi-port SRAM bitcell, simultaneous read and write operation in a single row or column degrades the read SNM, if conventional SRAM array organization is used. Therefore, alternative array organization is needed to address the issue of degraded read SNM. In this line, a word-oriented array organization is presented for a 2-port 6T bitcell to realize the high density SRAM array particularly suitable for future generation compact embedded systems realized as nanoscale SoCs. The presented 2-port 6T bitcell has following salient features as compared to existing 2-port (7T and standard 8T) SRAM bitcells:

- The 2-port 6T memory bitcell and its word-oriented array organization eliminates simultaneous read and write access disturbances due to column select function-ality in neighbouring bitcells or words.
- The poor read-noise margin and conflicting read-write problems are handled by isolating the read and write-ports to achieve higher stability margins.
- A misread (an erroneous read) or large bitline leakage current problem, existing in an isolated read-port 7T and standard 8T bitcells is significantly reduced by swapping the gate control signals of the read-port, with respect to 7T bitcell read-port.

- The process variation sensitivity analysis shows that the 2-port 6T bitcell design has significantly low process variation sensitivity as compared to existing SRAM bitcells, hence a better parametric yield can be expected.

4.2   Existing 2-Port SRAM Bitcells

The multi-port SRAM bitcell topologies are mainly used to increase the memory bandwidth in multi-core or parallel processors. However, exploiting the parallelism phenomenon in multi-port SRAMs in order to improve the bandwidth leads to certain design constraints. In this section, we will explore 2-port SRAM bitcell topologies and some design constraints or challenges that are inevitable in the nanoscale regime.

4.2.1   Standard 8T SRAM Bitcell

Figure 4.1 shows the standard 2-port 8T SRAM bitcell [83]. The 2-port 8T SRAM bitcell comprises of basic information storage unit (i.e. latch) and two read and write access ports. Each port has separate set of bitlines and a wordline for different read and write operations. This SRAM bitcell is identical to standard 6T SRAM bitcell with only difference of an extra read and write ports. In standard 8T, stability issues are quite similar to 1-port 6T bitcell such as conflicting read and write requirements

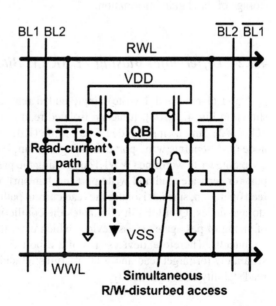

Fig. 4.1 Schematic diagram showing the simultaneous read/write-disturbed access (rise in the node voltage) and *dotted* read-current path (I_{read}) of a standard 2-port 8T SRAM bitcell [83]

of sizing the pass-gate devices, as a result large bitcell size. Hence, 2-port 8T SRAM bitcell has all the same problems that exists in standard 6T SRAM bitcell. Therefore, with the technology scaling, standard 8T has not been a popular choice of the SRAM designers in the nanometer regime because of conflicting read and write requirements that lead to poor noise margins and increased area overhead, as explained in the previous chapters.

Several new designs have been proposed in the recent past to address the nanometer regime issues. The prime concern in the SRAM design is the tradeoff among power, performance and area, while maintaining a higher degree of stability (or robustness). For example, in energy constrained applications such as sensor nodes and medical implants performance can be compromised with the power. Furthermore, in subthreshold SRAMs, noise margin (robustness) is the key design parameter and not the speed [112]. Handling of parametric yield loss due to stability issues or a poor read SNM, as shown in Fig. 4.3, and write-ability margin (WAM) simultaneously is a *challenging task*, because of tuning the cell ratio (β) for both the operations, while, maintaining an adequate I_{read}. It can observed from Fig. 4.3, that the read SNM of the standard 2-port 8T SRAM bitcell is almost $3\times$ less than the 2-port 6T and 7T SRAM bitcells.

Also the I_{read} (read-current path shown in dotted) has direct intervention with the data storage node and a strong relationship with the read SNM and read access time (i.e. performance). Hence, optimization of these parameters is not a trivial task. For instance, increasing the cell ratio will improve the read SNM and I_{read} but at the same time it will reduce the write-ability margin (WAM) and increase the bitcell size. Furthermore, the simultaneous read and write operations affect the contents of un-selected column bitcells and may flip the bitcells. If there is an insufficient SNM or WAM then the simultaneous read and write operations may lead to an un-faithful storage of the digital information.

4.2.2   Differential Biasing 7T SRAM Bitcell

A 2-port (1-read and 1-write) single-ended read and write 7T SRAM bitcell is shown in Fig. 4.2 [101]. It has an isolated read-port comprising of two transistors $M1_R$, $M2_R$, and a single-ended read bitline (RBL) to directly sense the data from node Q. A separate write port consisting of a single-ended write bitline (WBL) and a separate write wordline (WWL) controlling the pass-gate access device. A 1R/1W port (or separate read and write port mechanism) offers a static-noise-margin-free read operation, since, it isolates the read current path (shown in dotted) from the data storage nodes (Q or QB). It eliminates the conflicting read and write requirements of sizing of pass-gate access devices which exist in standard 1-port 6T and 2-port 8T bitcells. Therefore, devices size of a 2-port 7T SRAM bitcell can be optimize separately for target read and write margins to achieve a delicate balance between read stability and write-ability.

Fig. 4.2 Schematic diagram showing the simultaneous read/write-disturbed access (rise in the node voltage) and *dotted* read-current path (I_{read}) of an isolated read-port 7T SRAM bitcell [101]

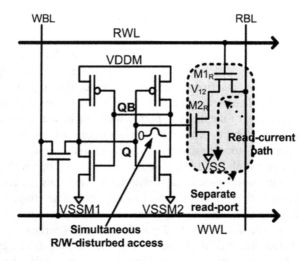

Fig. 4.3 The voltage transfer characteristics and SNM obtained from *butterfly curve* for the 2-port (dual-port) standard 8T, 7T and 6T SRAM bitcells

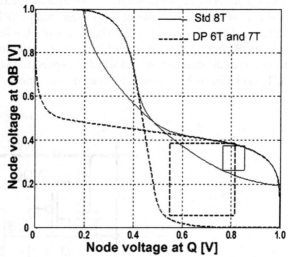

Separate read and write-ports provide about 3× better read SNM that cannot be achieved in standard 1-port 6T and 2-port 8T SRAM bitcells, as shown in Fig. 4.3. Maintaining a strong write-ability of logic '1' is difficult, specifically when a single-ended write bitline and a pass gate device are used [4, 44, 113]. Therefore, a data dependent differential VSSM (VSSM1 and VSSM2) biasing arrangement is proposed in this design to improve write-ability-margin (WAM). These VSSM lines are boosted (by $V_{SS} + \beta$) differentially depending on the input data. However, the use of differential biasing technique causes the undesirable loss of SNM at unselected bitcells in the write-selected column. Furthermore, generating and routing of these biasing increase the array area overhead and result in a complex layout and floor plan.

4.3 2-Port 6T SRAM Bitcell

In the nanoscale regime, for subthreshold SRAMs, noise margins or data stability
are the key concern [112]. In standard 1-port 6T or 2-port 8T bitcells read and
write noise margins significantly degraded due to poor voltage scalability of these
designs. Therefore, subthreshold operation in these design can only be achieved by
oversizing of the devices. Furthermore, in standard 1-port 6T or 2-port 7T and 8T
bitcells, stability problems also arise during a write or simultaneous read and write
operations to an unselected columns when the wordline is activated and bitlines are
asserted high [25, 101]. However, in order to cope with this type of stability loss,
column select functionality within the array must be eliminated or practical array
organization techniques need to explored. Therefore, modifications in the array
organization is just as important as modification in the SRAM bitcell itself.

Figure 4.4 shows the 2-port (1-read and 1-write) single-ended read and single-
ended write 6T SRAM bitcell. This bitcell consists of a cross-coupled inverter pair
(INV-1 and INV-2) and two single-ended separate read and write-ports. A separate
read-port comprises of a single-ended read bitline (RBL), transistor M1$_R$ and a
shared read-assist transistor M$_{RA}$. The transistor M1$_R$ separate's data storage nodes
(Q and QB) from the precharged RBL. The data is indirectly sensed from the bitline
according to the state of node QB which is connected to the write-port. If node QB is
at '1' the bitline is held high and sensed information is '1'. Similarly, if node QB is

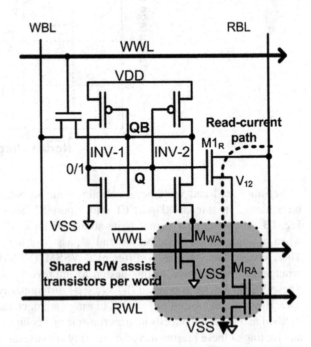

Fig. 4.4 Schematic diagram
of the 2-port 6T SRAM
bitcell with shared read and
write assist transistors per
word

at '0' the bitline is pulled down and sensed information is '0'. Separate read-current path (or read-port) also prevents node Q potential to rise when it holds '0'. The shared read-assist transistor M_{RA} is controlled by the read wordline (RWL) which activates the read-port.

The write-port consists of write wordline (WWL) and a write bitline (WBL) connected to node QB by a pass-gate device. The pass-gate device is controlled by the WWL. Since, write operation is performed by a single-ended bitline and it is assisted by a write-assist transistor (M_{WA}), shared per word. The write-assist transistor (M_{WA}) is controlled by the complement of write wordline ($\overline{WWL}$). The shaded transistors, as shown in Fig. 4.4 (M_{RA} and M_{WA}) are read and write-assist transistors, respectively, shared by all the bitcells of a word. The *unique features* of the 2-port 6T bitcell as compared to the previously proposed bitcells [83, 101] are as follows:

- The read bitline (RBL) is isolated with a single transistor while another (read-assist transistor, M_{RA}) transistor is shared among all the bitcells in a word. This arrangement provides a SNM-free read operation and a more area efficient bitcell, compared to standard 8T and differential VSSM 7T SRAM bitcells.
- Instead of having a dynamic or data dependent biasing scheme to improve write-ability, a write-assist transistor (M_{WA}), shared per word is used, to advance the WAM or to achieve a strong write-ability of logic '1' even at lower operating voltage levels (subthreshold).
- A non-interleaved array organization to facilitate the sharing of M_{RA}, M_{WA} and sub-wordline drivers, for eliminating the column select functionality within the array is presented. This helps in achieving both the SNM-free read operation and strong write-ability margin simultaneously, while eliminating the simultaneous read/write disturbance problems.
- Swapped control (gate) terminals of read-port transistors with respect to a 7T bitcell read-port, which minimizes the leakage current from the RBL by the unaccessed bitcells. This also helps in reduction of electrical loading effect on the data storage node due to reduced forward gate tunnelling current.

4.3.1 Array Organization

Figure 4.5 shows a 32-bit word-oriented SRAM array organization of the 2-port 6T bitcell, in order accomplish the target features, such as minimum area, better read and write noise margins. In this array organization, each word has more than 1-bit per word that is $n \geq 2$, where n is the number of bitcells (bits) in a word. Each word also has a sub-wordline driver to activate the local wordlines, and a set of read and write-assist transistors. To emphasize how the proposed array organization departs from the standard ones: in a standard SRAM array organization each word bitcells are interleaved (i.e. sandwiched). The interleaved and non-interleaved array organizations are explained in detail in Chap. 1. In a word-oriented SRAM array

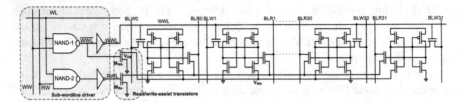

Fig. 4.5 A 32-bit word organization of the proposed 2-port 6T SRAM bitcell with shared read and write-assist transistors, and a sub-wordline driver to eliminate the simultaneous read/write disturbance problem

organization, all the bitcells of a word are kept together (i.e. non-interleaved), which facilitates the sharing of read and write-assist transistors. This results in each bitcell of a word having six transistors. Therefore, a word-oriented array organization with divided wordline is implemented, in which these shared transistors are activated vertically by sub-wordline drivers to read or write a word.

However, multi-divide word and bitline techniques are commonly used to reduce the length (charging and discharging capacitance) of wordlines and bitlines, or in other words to minimize the read/write delay for improving the array performance [47, 119]. The use of divided wordline and vertically activated sub-wordline drivers in this word-oriented array organization is a design strategy for achieving SNM-free read operation and strong write-ability margin simultaneously, while eliminating the simultaneous read/write disturbance or column select functionality problem within the array. It will increase an array area overhead, however, the main wordline drivers need to scale with their load, because they have to drive fewer sub-word line driver transistors, and can offset the area overhead incurred by the sub-wordline drivers. In this design sizing of read and write assist transistors play a significant role and they have direct impact on the performance, area and bitline leakage currents. The sizing issue of read and write transistors have been studied in detail in Chap. 3.

4.3.2   Read and Write Operations in 2-Port 6T Bitcell

The read operation of a 2-port 6T SRAM bitcell is carried out via a single-ended bitline (also known as data-line). Prior to a read operation, read bitline, RBL, is precharged to V_{DD}. After precharge, RBL is disconnected from the V_{DD} which is followed by the activation of read word line (RWL) to turn on the read assist transistor, $M2_R$, while the write wordline (WWL) of the bitcell is activated to low and its complement ($\overline{WWL}$) is set to high. High ($\overline{WWL}$) keeps the floating inverter to ground strongly during read operation in order to prevent the bitcell from inadvertent flipping. For reading '0', read bitline (RBL) has to discharge through the read-port (i.e. from $M1_R$ and $M2_R$), while for reading '1', read bitline has to remain at precharged level ($\sim V_{dd}$) because transistor M2$_R$ is turned off. As a result, reading '1' or '0' is directly sensed from the precharged RBL. Therefore, in either case reading '1' or '0', storage nodes are isolated from the read current path, hence, it significantly enhances the data stability during read cycle.

The write operation of this 2-port 6T SRAM bitcell is similar to the single ended 6T SRAM bitcell, as presented in Chap. 3. However, a dedicated write bitline (WBL), and write wordlines (WWL and ($\overline{WWL}$)) are employed. The stability margins read SNM and WNM analysis presented in Chap. 2 for single ended 6T SRAM bitcell also hold true for 2-port 6T bitcell. Hence both designs employ the same stability enhancement techniques, such as isolated read-port and weakening of regenerative mechanisms.

4.4 Reconfigured Read-Port of a 2-Port 6T Bitcell

The leakage currents drawn by the un-accessed bitcells from pre-charged bitlines restrict the number of bitcells per bitline to avoid the erroneous or misread operation. This mainly happens when sense amplifier fails to distinguish between pre-charged and discharged bitlines. This problem exists in both standard 6T and 8T SRAM bitcells and in isolated read-port SRAM bitcell designs. Since each bitcell has two transistors connecting the bitlines to ground. However, bitline leakage current problem is more severe in standard 6T or 8T SRAM bitcells where small differential voltage has to be sensed by the sense amplifier. Another difference between standard 6T and isolated read-port 6T SRAM bitcell is that the leakage current from the bitline is data dependent in isolated read-port SRAM bitcells, while standard 6T SRAM bitcell is symmetric in nature and one of the bitline always has this leakage current path. The aggregated leakage current from all the unaccessed bitcells can pull-down the bitline, even if the accessed-bitcell stored information does not need to do so. In other words, if the aggregated bitline leakage current exceeds the actual read-current, correct data sensing becomes impossible or an erroneous read may occur. However, some attempts have been made to minimize the bitline leakage current problem at the cost of increased number of transistors in the read-port and extra peripheral circuitry such as read buffer footer [18, 108].

For instance, in an 7T SRAM design, read-ports of all the bitcells in a column are connected to a single read bitline (RBL) and the information is directly sensed from the RBL. Prior to each read operation that is selecting a particular row (RWL =1), RBL must be precharged so that the read-port can sink the precharged RBL to a sensing level ($0.5 \times V_{DD}$) to sense the bitcell information from the RBL, if the node Q is at 0V, otherwise it holds RBL to $\sim V_{DD}$. During read operation, all the bitcells in the column are unaccessed except one, which is corresponding to selected read wordline (RWL). If the aggregated leakage current drawn by these unaccessed bitcell's read-ports from the RBL is higher than the I_{read} of the read bitcell, than it may wrongly pull down the precharged RBL to a sensing level results a misread operation. Thus, the read-port configuration is very decisive in an isolated read-port

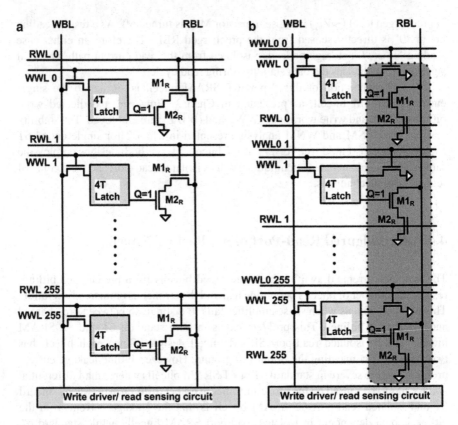

Fig. 4.6 Schematic diagrams of a column showing the 256 bitcells per read bitline (**a**) bitcells with standard isolated read-port (**b**) bitcells with re-configured isolated read-port

SRAM bitcell designs, as it also forms the critical read access path. Figure 4.6a, b, respectively, show a column shared by 256 bitcells with their standard (7T) and the re-configured read-ports (6T) connected to a single RBL.

The subthreshold leakage current drawn from the RBL is totally dependent on the state of the unaccessed bitcells and their read-ports, because of asymmetric nature of read operation. The worst case is when all the unaccessed bitcells hold '1' at node Q, as shown in Fig. 4.6a, b. Each read-port configuration in an 7T/8T bitcell is comprised of two transistors $M1_R$ and $M2_R$ and these transistors are controlled by two control signals *RWL and Q*, respectively. The only difference between these two configurations of the read-ports is the swapped control signals, that is in re-configured read-port transistors $M1_R$ and $M2_R$ are controlled by signals *Q and RWL*, respectively. However, read and write operation in both configurations take place in a similar fashion. An important observation which makes these two similar configurations significantly different, not only by swapping their control (gate terminal) signals but also by the leakage current drawn from the read bitline

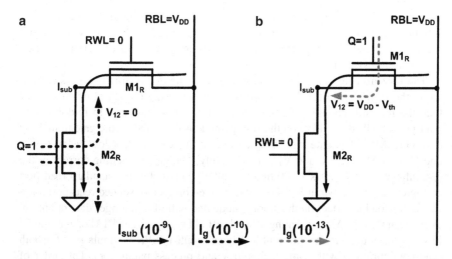

Fig. 4.7 (**a**) Standard isolated read-port for 7T/8T SRAM bitcell (**b**) re-configured isolated read-port of 6T SRAM bitcell

(RBL) and the forward gate tunnelling current drawn from the data storage node (Q) which causes in a electrical loading effect on the data storage node Q.

Let us, examine both standard and re-configured isolated read-ports individually when a bitcell is read for clarification, as shown in Fig. 4.7. We assume that the read bitlines are precharged to V_{DD} (RBL = 1), the read wordlines (except one) are asserted low (RWL = 0) and the data storage node Q of all the bitcells in a column hold '1' (Q = 1). Under these conditions, for a standard read-port configuration gate terminals of transistor $M1_R$ and $M2_R$ are connected to RWL (=0) and node Q (=1), respectively, as shown in Fig. 4.7a. Therefore, drain to source (V_{DS}) voltage of $M2_R$ (i.e. V_{12}) and $M1_R$ is 0V and V_{DD} respectively, as shown in Fig. 4.7a. The subthreshold leakage current (I_{sub}) drawn from the RBL, which is tied to the drain of $M1_R$ is dependent on the V_{DS} of $M1_R$ (i.e. V_{DD}). The forward gate tunnelling current in the standard read-port will occur significantly at transistor $M2_R$ because the gate tunnelling current occurs maximum when the V_{GS} is at V_{DD} and $V_{DS} = V_{SS} = 0V$ for an NMOS device [67]. As a result loading effect on the bitcell will be exhibited by $M2_R$, hence, a standard read-port experiences a maximum loading on the storage node (Q).

Figure 4.7b shows the re-configured read-port, where gate terminals of transistor $M1_R$ and $M2_R$ are connected to node Q (=1) and RWL (=0), respectively. In this configuration, subthreshold leakage current flows from the transistor $M2_R$ while drain to source voltage of $M2_R$ (i.e. V_{12}) is reduced by a factor of subthreshold voltage drop which is ($V_{DD} - V_{TH}$) with the body effect. As the subthreshold leakage current (I_{sub}) from the RBL is dependent on the drain to source voltage of $M2_R$ i.e. ($V_{DD} - V_{TH}$). The drain to source voltage of the leaky (subthreshold) transistor $M2_R$ is reduced by a factor of V_{TH}/V_{DD} compared to $M1_R$ of standard read-port, as shown in Fig. 4.7a. Thus, the re-configured read-port can *significantly reduce* the subthreshold leakage current from the RBL. Thereby, reduction in the subthreshold leakage from the RBL helps in avoiding the erroneous or misread operation caused by the un-accessed bitcells.

The forward gate tunnelling current in a re-configured read-port occurs at $M1_R$ while the reverse gate tunnelling current occurs in $M2_R$ (which is much smaller than the forward tunnelling current, and hence can be ignored). In re-configured read-port, the loading effect on the storage node is exhibited by the gate tunnelling current of $M1_R$. The gate tunnelling current in this case is very low because V_{GS} and V_{GD} of $M1_R$ is 0 V and V_{TH}, respectively. The gate tunnelling current in the re-configured read-port is 1,000 times smaller compared to the standard read-port configuration, as shown in Fig. 4.7. Thus, the re-configured isolated read-port has a negligible loading effect on the storage node that will offer a longer data retention, even when the SRAM is operating at subthreshold voltages. The PTM 65 nm, model [88] was used to compute different leakage currents in SPICE simulations for both read-configurations with same technology and process parameters. The order of magnitude of the different leakage currents (as shown in Fig. 4.7) are based on $V_{DD} = 1.0$ V.

4.4.1   RBL Leakage and Gate Tunnelling Currents

A large subthreshold leakage current drawn from the RBL due to unaccessed bitcells limits the number of bitcells per RBL. Also large subthreshold leakage current makes conventional data sensing impractical [18]. This is an issue in both standard bitcell structure (6T) and in an isolated read-port bitcell structure (7T/8T), because both have equal number of transistors in the read current path (i.e. two). In [108], a read-foot buffer shared among the bitcells per word is employed for reduction in the RBL leakage current. This arrangement reduces the bitline leakage and avoids the misread operation and the limitation of bitcells per bitline. However, it needs considerable tradeoffs in floorplanning or silicon overhead. In contrast, re-configured read-port enables reuse of standard 6T read-port 7T SRAM bitcell without any tradeoffs in floorplanning or silicon overhead.

The forward gate tunnelling and subthreshold leakage currents of a 32×256 module with standard and re-configured read-ports are shown in Fig. 4.8. In the re-configured read-port, forward gate tunnelling current is 93% less than the standard read-port. The subthreshold leakage current in the re-configured read-port is reduced by 54% as compared to standard read-port. A significant reduction in the gate tunnelling and subthreshold leakage currents is achieved because RWL is controlling the $M2_R$ in re-configured read-port. It is observed that the re-configuration of the read-port can reduce the subthreshold leakage from RBL and gate leakage currents significantly. Thus, the use of re-configured read-port configuration in 7T/8T SRAM bitcell will help in avoiding the misread operation due to false pull down of RBL and restricting the number of bitcells per RBL. Also it will reduce the overall leakage and active power consumption in the SRAM design.

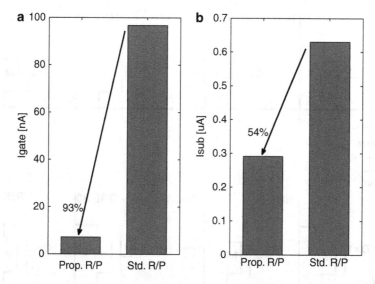

Fig. 4.8 A 7T/8T 32×256 SRAM module implemented with standard and re-configured read-ports(R/P) (**a**) gate tunnelling leakage current (**b**) subthreshold leakage current

Two worst case scenarios of the RBLs discharging pattern have been studied to demonstrate the efficacy of the 6T bitcell and re-configured read-port and its array organization during read operation. For SPICE simulations, a 32 bit word and 256 bitcell per column (or per RBL) is considered. In order to understand, how the un-accessed bitcells in a worst case will affect the performance mainly due to excess leakage current from the bitline via read-assist transistor, when a word is a read. This leakage current also include the leakage from the read-assist line shared by all the bitcells (32) read-buffer's in a word.

4.4.2   Read Bitline Leakage Scenario-1

Figure 4.9a shows a possible worst case scenario, where all the bitcells in an accessed word hold '0' (i.e. Q = 0), while remaining un-accessed bitcells hold '1'. Under these circumstance RBLs must remain at precharged level except small residual droop, in order to avoid misread operation.

Also under the sequential operation that is when changing (writing) the state of a node Q from '0' to '1', all RBLs are floating and the read-assist transistor (M_{RA}) is in cut-off. Therefore, the leakage current from RBLs under sequential or during the successful write ($0 \rightarrow 1$ or $1 \rightarrow 0$) state is not significant as compared to the hold (or idle) state when Q = 1 (worst case). The hold state leakage current for different word sizes and bitcells per bitline is exhaustively studied in the Chap. 3. An optimal size of the read assist transistor for target specifications of leakage current and delay can also be derived from those simulation results.

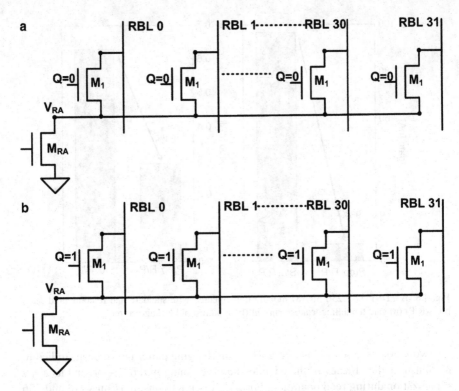

Fig. 4.9 (a) Read bitline leakage scenario-1 (b) read bitline leakage scenario-2

The transient SPICE simulations for RBL discharge pattern using a 32×256 module of 7T/8T bitcells with standard and re-configured read-ports are shown in Fig. 4.10. Thirty two-bitcells per word and 256 bitcells per RBL are used to demonstrate the bitline discharging pattern due to leakage in worst case by the unaccessed bitcells. Figure 4.10a shows that the unaccessed bitcells can pull down the RBL to a sensing level of $0.5 \times V_{DD}$ in a time of $10\,\mu s$, which results a misread operation in 7T/8T SRAM with standard read-ports module only when the sampling window is large. However, re-configuration of read-ports in an 7T/8T SRAM module reduces the leakage from the RBL, from unaccessed bitcells, and keeps RBL high, as shown dotted in Fig. 4.10a. That is how reduction in the leakage current from the RBL helps in avoiding the misread operation.

4.4.3 Read Bitline Leakage Scenario-2

In order to read a word from the module, under scenario-2, as shown Fig. 4.9b, the RBLs should discharge as fast as possible in worst case when all the bitcells

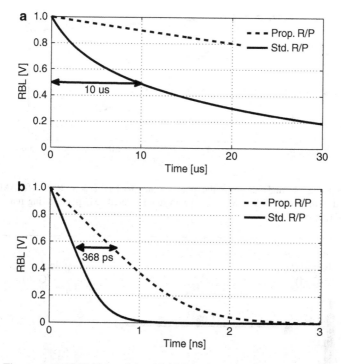

Fig. 4.10 The standard 7T SRAM module with 256 bitcells per bitline implemented with standard and re-configured read-port(R/P) (**a**) bitline leakage discharging pattern under hold state (**b**) bitline discharging pattern for read '0'

in a word hold '1' (i.e. Q = 1), while remaining bitcells in the column hold '0'. Figure 4.10b shows the RBL discharging pattern in worst case read operation for iso-read current with standard read-ports and re-configured read-ports. One can observe that the re-configured read-port module is slower by 368 ps compared to the standard read-port module because of the stacking phenomena [67]. Furthermore, there is a read assist line shared among the read-ports of all the 32-bits at node V_{RA}, which has certain capacitance leads to increase the read access time. Since the node capacitance is equally shared by all the RBLs in a word and it is quite smaller compared to the bitline capacitance. Therefore, there is little impact on the performance of the SRAM module with re-configured read-port. This delay can be further optimized by sizing the M2$_R$ of the re-configured read-ports and a similar performance can be achieved.

Figure 4.11 shows the SPICE transient simulations of a 32×256 SRAM module during read access. It is shown that the RBL is correctly pulled low by the accessed bitcell when it holds '0' (i.e. Q = 0). Also it remains high when the accessed bitcell hold '1' and unaccessed bitcells hold '0'. It can be observed that RBL is not erroneously pulled low by the leakage currents of the un-accessed 255 bitcells connected to the same RBL. However a small 10% residual droop can be observed

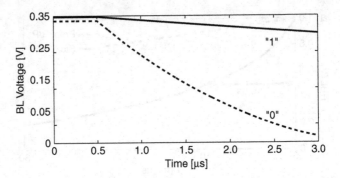

Fig. 4.11 Sensing '1' and '0' from the read bitline (RBL) of a 32×256 SRAM module implemented with 2-port 6T SRAM bitcell (with re-configured read-port) during read access in worst case scenario

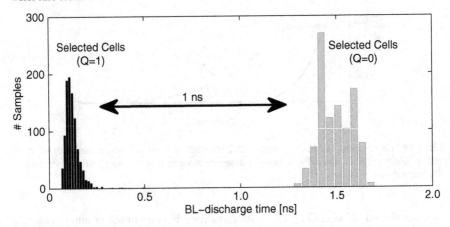

Fig. 4.12 In 2-port 6T SRAM bitcell with re-configured read-port read bitline (RBL) discharge time distribution for storage node Q is '1' and '0'

while reading '1', due to the gate-oxide and junction leakage from the read assist transistors. In 8T [108], an erroneously pull down problem by the leakage currents from the un-accessed bitcells is solved by the read-buffer foot, which is shared among the bitcells of a word and leads to extra area overhead.

Furthermore, in order to illustrate that the variation in RBL leakage current will not cause any misread or erroneous reads. In other words, it will provide a distinguishable bitcell read current from the leakage current, transient SPICE simulations are performed. The Monte Carlo simulations for RBL discharge time was compared for read '1' and '0' operations. The distribution of RBL discharge time is measured from V_{DD} to $0.5 \times V_{DD}$ when the bitcell storage node Q is 'high' and 'low'. The 32 bitcells per word and 256 bitcells per RBL was used in the compiled module. The distribution of RBL discharge time with 255 unaccessed bitcells for Q is '1' and '0' is shown in Fig. 4.12. It can be seen that the clearance of RBL discharge time between the slowest bitcell for $Q = 1$ and the fastest bitcell for $Q = 0$ is more than 1 ns, so enough timing margin to latch the data exactly.

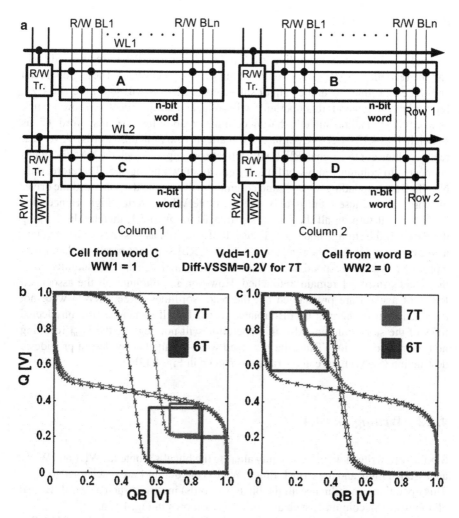

Fig. 4.13 (**a**) A schematic diagram for illustrating the simultaneous read and write access issues in the word-oriented array organization with 2-port 6T SRAM bitcells, and (**b**) and (**c**) *butterfly curves* for SNM comparison when a bitcell of word 'A' is read and written

4.5 Simultaneous Read/Write Access in 2-Port 6T-SRAM

Figure 4.13a shows the schematic diagram of a 2-port 6T SRAM bitcell memory module, with word-oriented array organization in which four n-bit words (viz A, B, C and D) are arranged in two-rows and two-columns. In order to demonstrate how simultaneous read and write accesses influence the state of the neighbouring bits or words. The butterfly curves shown in Fig. 4.13b, c are used for measuring the degree of disturbance in terms of static noise margin (SNM).

4.5.1   Reading Word A

To read word 'A', global wordline WL1 and read wordline RW1 are asserted high, and the read bitlines (RBLs) of column 1 are precharged to V_{DD}. These global read and write wordlines, with the help of sub-wordline drivers will select the word 'A', for reading. In general, this operation will influence all the bitcells in row 1, such as word 'B' and all the bitcells sharing column 1 read bitlines, such as word 'C', as shown in Fig. 4.13a. Let us examine, how it affects row 1 bitcells (words). A vertical read wordline control signal, RW1, will activate one input of each the NAND-2 gates connected to column 1 sub-wordline drivers, as shown in Fig. 4.5. Therefore, only NAND-2 corresponding to row 1 and column 1 will activate the sub-wordline driver of word 'A', because there is only one wordline WL1 asserted high. Hence, this NAND-2 will turn on all the read-assist transistors of row 1, thereby discharging the RBLs, if the storage node Q = 1, else, RBLs remain as it is. Also when reading a word from column 1, the remaining columns' RBLs are not precharged and sub-wordline drivers correspond to these columns are inactive. Thus, the stability of all the bitcells in row 1 remain untouched. However, in column 1 all the associated RBLs of a word are precharged but the read wordlines (except RW1) were not activated, hence, there is no disturbance to the bitcell content of the unselected rows of the same column. Also this operation will not degrade the I_{read} resulting non-misread operation. Thus, the proposed word-oriented array design provides a destruction (SNM) free read operation as shown in Fig. 4.13b.

4.5.2   Writing Word A

Similarly to write in word 'A' (mainly altering the bitcells' content), WL1 and WW1 are asserted high, and the write bitlines (WBLs) of column 1 are precharged to V_{DD}. This operation can influence all the bitcells (words) in row 1, such as word 'B' and all the words in column 1, such as word 'C', as shown in Fig. 4.13a.

   Write operation in a selected word 'A', will only take place when NAND-1 corresponding to row 1 and column 1 (see Fig. 4.5) will activate the sub-wordlines driver of word 'A'. This is because the NAND-1 has to drive the local sub-wordlines to turn-on all the access transistors connecting the WBLs and turning-off the write-assist transistors of all the bitcells in a word. Also, when writing into column 1, remaining WBLs (except column 1 WBLs) were not precharged and these WBLs will not be connected to bitcell data storage node by the access devices. Thus, the stability of all the cells in row 1 words remain untouched. For column 1, all the WBLs associated with word 'A' were precharged but the write wordlines (except WL1) were not activated, thereby the write access device of remaining bitcells sharing the same column are in cut-off, hence, there is no significant influence to the bitcell content of the unselected words of the same column (see Fig. 4.13b).

For a comparative view, study of the write disturbance in the bitcells of word B, C and D is presented. The SNM metric was obtained from the butterfly curves to study the disturbance in the bitcells around the written word A (WL1 = 1 and WW1 = 1). The butterfly curves of a bitcell from the word B are obtained by keeping the WL1 = 1 and WW2 = 0 as shown in Fig. 4.13b. The SNM of a 6T bitcell from the word B is 61% higher than the 7T bitcell [101]. The 7T bitcell SNM is disturbed due to voltage division effect between an access and a NMOS pull down transistor and also due to differential bias arrangement for write operation. Similarly, for a bitcell from word C, the read SNM is 53% better than the 7T bitcell because the use of differential VSSM disturbance.

4.5.3 Simultaneous R/W Word A and C

Simultaneous read and write operations in the previously proposed schemes of 2-port bitcell designs [23, 101, 103], possess some challenges such as maintaining sufficient read SNM, WAM and I_{read}. Reduction in read SNM and increase in I_{read} is mainly caused when the storage node Q voltage rises. The increase in the node Q voltage is due to the voltage division effect when a simultaneous read and write operations occur. An increase in I_{read} may cause misread operation due to increased RBL leakage, while reduction in SNM may flip the data storage node content. In the 2-port 6T bitcell, a simultaneous read and write operation of a word from the same column, such as reading a word 'A' and writing a word 'C' is illustrated using Fig. 4.13a. In a 6T bitcell, storage node Q will not be disturbed due to the use word-oriented array design with sub-wordline drivers and modified read-port configuration. These features help the 6T bitcell to keep the SNM free read and write operation, thereby any misreading does not occur, and also there is no SNM reduction.

4.6 SRAM Process Variation Sensitivity

Systematic and random variation in process, supply voltage and temperature (PVT) have become a major challenge in high density SRAM. Process variation is mainly caused by the difficulties in the precise control of lithography and inherent random process, thus causing the line edge roughness. With the continuing reduction of the number of dopant atoms in the channel between source and drain in a MOSFET, random dopant-density fluctuation causes variation in the transistor characteristics. Temperature variation can lead to hot-spots and points of reliability failure and has an exponential relationship with leakage current which can cause further exacerbation of power. These variations translate into uncertainties in the circuit performance metrics.

In general, process variations can be divided into the following two broad categories: inter-die or die-to-die (D2D) variations and intra-die or within-die (WID) variations. Inter-die variations are the variation from die to die and they affect all the devices on the same chip in the same way. For example, making the transistor gate lengths of on the same chip all larger or smaller. Intra-die variations correspond to variability within a single chip and may affect different devices differently on the same chip. For example, making some devices have smaller transistor gate length and others larger transistor gate length. In addition, WID variations exhibits spatial correlation, that is, devices sitting next to each other are more likely to have the similar characteristics than those sitting far away. Mathematically, inter-die variations can be regarded as a special use of intra-die variations with correlation value of one [22].

Process variation sensitivity of standard 6T and 2-port 6T SRAM bitcells is studied to determine the robustness in terms of SNM sensitivity. The SNM sensitivity analysis for known device parameter variations such as W, L and V_{TH} is performed. Small variations in these parameters were made to identify the SNM sensitivity. It helps in determining which parameter in which device will affect the SNM sensitivity and how much variations a design can tolerate. Therefore, it can be a good measure of parametric yield in SRAM bitcells. Lower the SNM sensitivity to process variations better the SRAM bitcell. The SNM sensitivity to a device parameter $x(W, L$ or $V_{TH})$ on device i is defined as the per unit change in SNM (Δ SNM) to per unit change in parameter (Δx), that is $\frac{\Delta SNM}{\Delta x_i}$. These sensitivities are obtained from the SPICE simulations by small variations in x_i, as shown in Figs. 4.14 and 4.15 for a standard 6T SRAM and the 2-port 6T SRAM bitcells, respectively at $V_{DD} = 1.0$ V.

Small variation in different design and process parameters such as channel length L, channel width W and threshold voltage V_{TH} is considered to see the variation in read SNM of different SRAM bitcell designs. It can be observed from Fig. 4.14 that the variation in channel length (L) and threshold voltage (V_{TH}) of different transistors of a standard 6T SRAM bitcell has large impact on the SNM. However, variation in channel width (W) has least impact of the read SNM. Similarly, in 2-port 6T SRAM bitcell, variation in channel length (L) and threshold voltage (V_{TH}) has large impact on SNM as compared to variation in channel width (W), as shown in Fig. 4.15.

Simulation results show a linear relationship between ΔSNM and small variations in x_i. Hence, SNM sensitivity is the gradient of these straight lines, higher the gradient higher the sensitivity. In both the SRAMs, SNM is more sensitive to variations in L followed by V_{TH} and W. It indicates the pronounced short channel effect of drain-induced barrier lowering(DIBL), which significantly deteriorates SNM, since it reduces inverter gain at high V_{DD}. SNM sensitivity in standard SRAM bitcell due to variations in L of pull down devices (M_2 and M_4) is higher than the pull up devices (M_1 and M_3) and followed by the access devices (M_5 and M_6), as shown in Fig. 4.16. The trend is well expected, since, pull down devices dominate in controlling the SNM. The high SNM sensitivity to standard SRAM bitcell is mainly due to opposite nature of gradient between a pair of pull down devices (M_2 and M_4), pull up devices (M_1 and M_3) and access devices (M_5 and M_6). For instance, the gradients of pull down devices for variations in L are 1.43 and -1.2, respectively,

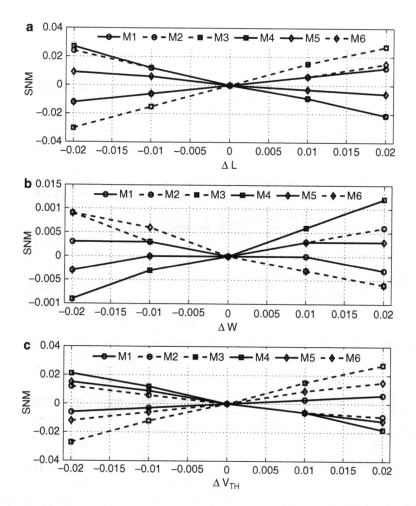

Fig. 4.14 Standard 6T SRAM read SNM variations obtained from the *butterfly curves* by varying the device parameters (**a**) small change in L, (**b**) small change in W, and (**c**) small change in V_{TH}

as shown in Fig. 4.16a. Consequently, standard 6T SRAM bitcell is more sensitive to process variations as compared to separate read-port 6T SRAM bitcell.

Process variation sensitivity results of the separate read-port 6T SRAM bitcell under small variations in device parameters show the similar trend, that is, channel length and threshold voltage variation has large impact as compared to channel width. However, low SNM sensitivity is observed as compared standard 6T SRAM bitcell as shown in Fig. 4.15. The short channel effect is also pronounced here, and makes higher SNM sensitivity to channel length L. An asymmetric nature and separate read and write ports make the proposed SRAM design less sensitive to read SNM, as indicated in Fig. 4.16a. For instance, the gradients of pull down devices

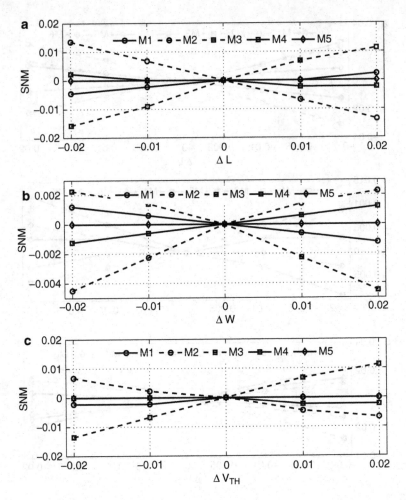

Fig. 4.15 Proposed 6T SRAM read SNM variations obtained from the *butterfly curves* by varying the device parameters (**a**) small change in L, (**b**) small change in W and (**c**) small change in V_{TH}

(M_2 and M_4) for variations in L are 53% and 91% less compared to standard 6T SRAM bitcell as shown in Fig. 4.16a. As a result, proposed SRAM is more robust against process variations and may provide better parametric yield. However, this sensitivity analysis is also true for the 6T SRAM bitcell design discussed in Chap. 3.

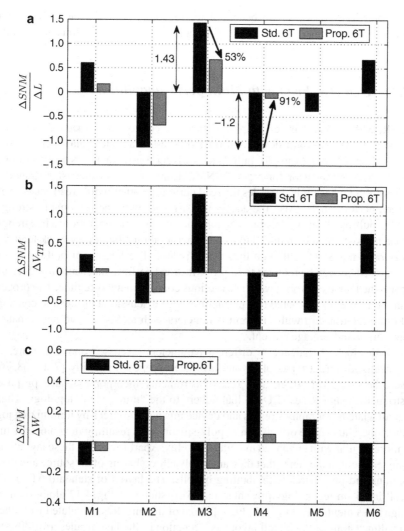

Fig. 4.16 The gradient defined as per unit change in SNM to per unit change in the device parameter (**a**) L, (**b**) V_{TH} and (**c**) W, at $V_{DD} = 1.0$ V for standard 6T SRAM and separate read-port (Prop.) 6T SRAM bitcell

4.7 Area, Power and Performance of the 2-Port SRAM Bitcells

Area, power, and performance are three key the metrics apart from stability and process variability tolerance in the SRAM design to identify a potential SRAM design for specific applications. These metrics have significant importance when multi-port SRAM designs have been targeted for sophisticated applications such as pipelined or parallelism in embedded multimedia and communication applications.

4.7.1 Area Overhead with Multi-port Capabilities

The SRAM bitcell area is very important from the economic perspective and sizing trade-off of different (pull-down, pull-up and access) transistors play a crucial role for functionality. For example, pull-down (driver) transistor needs to be stronger than the access transistor for a good SNM. However, access transistor cannot be made too small since this degrades the read current, in other words read access time (or read performance). Also the access transistor needs to be reasonably stronger than the pull-up (load) transistor to enable successful write operation. The strength of load transistor can be reduced to improve write-ability, but a very weak load will deteriorates the SNM, although the impact is less. The lengths of pull-down and access transistor can be reduced to improve the read performance, but this adversely impacts the leakage current, which is a serious concern in nano-regime. The process variation further exacerbated this problem. Above discussion clarifies the concerns and clsoe relationship with different parameters such as SNM, read performance, write-ability and leakage current.

In order to keep balance in different parameters discussed above and area of the bitcells, different types of layout have been proposed recently [7, 15, 68, 76]. Researchers at IBM, Intel, and Texas Instruments Incorporated have proposed Restrictive Design Rules (RDRs) that adhere to the "thin bitcell" topology. These rules alleviate lithography stresses and device mismatch sources by minimizing jogs in the poly. Thus single orientation of poly-silicon gates resulting in geometries that are more regular with enhanced manufacturability. Apart from that, these layouts are highly efficient in the sense that they amenably allow sharing of all source and drain junctions and poly wires with abutting bitcells. The layout of standard 6T SRAM bitcell in 65 nm technology presented in [15] is shown in Fig. 4.17 along with the design rules mentioned in [7, 15, 39, 68]. Area of the bitcell is calculated from the x and y dimensions of the bitcell layout as a function of the layout rules as follows:

$$Area = x_{dim} \times y_{dim}$$

$$x_{dim} = 2 \times max(x1, x2)$$

$$y_{dim} = 2 \times max(y1, y2)$$

$$x1 = \left(\frac{1}{2}\right)(PP) + W_{ld} + PN + W_{drv} + PoG + \left(\frac{1}{2}\right)(PoPo)$$

$$x2 = \left(\frac{1}{2}\right)(PP) + W_{ld} + PN + W_{ax} + PoG + \left(\frac{1}{2}\right)(CW)$$

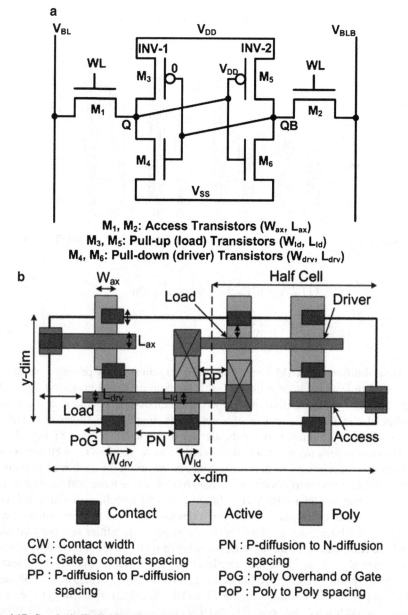

a

V_{BL} V_{DD} V_{BLB}

WL   INV-1   INV-2   WL

M_3 0 V_{DD} 0 M_5

M_1 M_2

Q   QB

M_4 M_6

V_{SS}

M_1, M_2: Access Transistors (W_{ax}, L_{ax})
M_3, M_5: Pull-up (load) Transistors (W_{ld}, L_{ld})
M_4, M_6: Pull-down (driver) Transistors (W_{drv}, L_{drv})

b

W_{ax} Half Cell

Load   Driver

L_{ax}

y-dim

PP

L_{drv} L_{ld}

Load   Access

PoG   PN

W_{drv} W_{ld}

x-dim

■ Contact   ■ Active   ■ Poly

CW : Contact width
GC : Gate to contact spacing
PP : P-diffusion to P-diffusion
     spacing

PN : P-diffusion to N-diffusion
     spacing
PoG : Poly Overhand of Gate
PoP : Poly to Poly spacing

Fig. 4.17 Standard 6T CMOS SRAM bitcell and its layout. (**a**) Standard 6T SRAM bitcell. (**b**) SRAM bitcell layout with Restrictive Design Rules (RDRs)

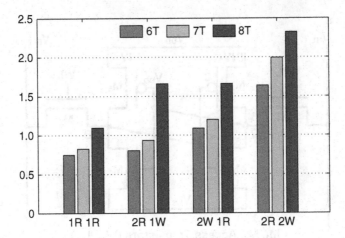

Fig. 4.18 Comparison of area overhead for multi-port capabilities in 6T, 7T and 8T bitcells

$$y1 = 2\left[\left(\frac{1}{2}\right)(CW) + 2(GC) + L_{ld}\right] + CW$$

$$y2 = 2\left[\left(\frac{1}{2}\right)(CW) + 2(GC)\right] + L_{drv} + L_{ax} + CW. \qquad (4.1)$$

Area of different SRAM bitcells and their additional ports provided for multi-port capabilities have been estimated with the help of RDRs presented above. In general, multi-port capabilities in the SRAM bitcell designs quadratically increases the bitcell size with the number of access ports. Figure 4.18 shows the trend of area overhead with multi-port capabilities for different 6T, 7T and 8T bitcells for a 45 nm technology node. An additional read or write port in 8T bitcell needs two bitlines, two access devices and a wordline. However, in 7T each read port costs of two isolated read-port devices and a read bitline, while write port needs a single write wordline and an access device. In the 2-port 6T bitcell, each additional read or write port will cost a read or write assist device and a read or write bitline. Thus, the 2-port 6T bitcell design provides the multi-port capabilities at a reduced area overhead compared to 7T and 8T bitcells, either providing a read port or a write port. An analytical analysis of the area overhead of 6T, 7T and 8T bitcells is presented here. The 2-port (1R and 1W) 6T bitcell area is $0.748\,\mu m^2$ i.e. $(0.55\,\mu m \times 1.36\,\mu m)$, which is 9% and 31% lower than the 7T and 8T bitcells, respectively. In order to provide an additional read port (2R and 1W) the area overhead for the 8T bitcell goes 100% higher than the 6T and 77% higher than the 7T bitcell. However, the area overhead for 6T bitcell is 16% lower as compared to 7T bitcell. Similarly, area overhead cost for providing 2R and 2W ports in a in 6T bitcell is 22% and 42% less compared to 7T and 8T bitcells, respectively.

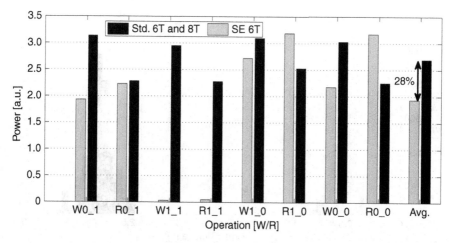

Fig. 4.19 Active power pattern for different read/write operations of SE and standard 6T SRAM bitcells

4.7.2   Power Dissipation

Figure 4.19 compares active power in the 6T, 7T and 8T bitcells for different read/write operations. As 6T and 7T bitcells are asymmetric in nature, hence, their active power consumption pattern is also asymmetric. In Fig. 4.19, operation W0_1 stands for writing '1' into the bitcell while its original content was '0'. Similarly, R1_0 stands for reading '0' from the bitcell, while its previous output was '1'. For operations W1_1 and R1_1 the active power of 6T/7T bitcells is very low as compared to 8T bitcell, because both the operations are performed without discharging the bitline of the 6T/7T bitcells. Under such operations precharged bitline can be used for future read/write operation. Alternatively, in 8T bitcell one bitline has to discharge during these operations. However, the active power for operations R1_0 and R0_0 in 6T/7T bitcells is 21% and 29% higher than the 8T bitcell. The average active power under different read/write operations of the 6T or 7T SRAM cell is 28% lower than the 8T bitcell (Fig. 4.19).

4.7.3   Performance

For the target applications such as video-processing, high read access multi-port SRAM is strongly recommended since the read operation occurs more repeatedly than the write operation in video codec. For instance, in video codec once video frames are written in memory, several search algorithms have to read the data many times for decoding those frames. Figure 4.20 compares the distribution of the read access time of 6T, 7T and 8T bitcells. The read access time distribution was obtained

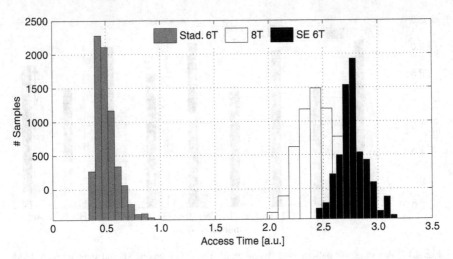

Fig. 4.20 Distribution of read access time of 6T, 7T and 8T bitcells

by Monte Carlo simulations. Each bitcell was simulated under 3σ random variations in threshold voltage of each transistor. For the proposed 6T and differential VSSM 7T read access time was calculated when the read wordline (RWL) rises to $0.5 \times V_{DD}$ to a time when the output of the sense amplifier (read buffer) is reached to $0.5 \times V_{DD}$. Similarly, in 8T read access time was defined as the time between the RWL rises to $0.5 \times V_{DD}$ to a time when we got the expected differential voltage of bitlines (50) mV. The mean read access time of 6T and 7T bitcells is very close that is 2.76 ns and 2.48 ns, respectively. Read access time of the proposed 6T bitcell is 10% higher than that of the 7T bitcell because of the modified read-port or in other words, stacking phenomena in the read-port slow down the read performance of the 6T bitcell. However, mean read access time of 8T bitcell is significantly lower compared to 6T and 7T bitcells. Hence, 8T bitcell achieves the high performance. Performance of the 6T can be achieved equivalent to an 8T by optimizing the size of read-assist transistor, however, it may lead to an increase in area overhead.

In multiport SRAM designs, overall performance in terms of throughput can be greatly improved by providing the extra number of ports per bitcell. Generally, adding the number of ports costs extra silicon overhead, leakage and dynamic power dissipation. Therefore, throughput of the proposed design can be improved by adding the number ports to equalize the performance at the reduced area overhead, leakage and dynamic power dissipation as compared to 8T bitcell design.

4.8 Summary

In this chapter, study of different 2-port SRAM bitcell with multi-port capabilities is presented. The poor data stability, read and write disturbances and simultaneous read and write conflicts issues of this bitcell are addressed. The robustness, area overhead, power and performance of different bitcells are compared. The 2-port 6T bitcell has better static noise margin compared to 7T bitcell under write disturb conditions. Also, 6T bitcell provides SNM free read operation while in 7T and 8T designs SNM degrades during read operations and simultaneous read and write access. The bitline leakage current by the unaccessed bitcells is reduced in the 6T due to re-configuration of the isolated read-port, results, no misread operation. The area overhead in the 2-port 6T bitcell for providing the multi-port capabilities such as additional read and write ports is lower than the 7T and 8T bitcells. Hence, the 2-port 6T bitcell design has significant potential for the multimedia and communication applications for nanoscale and other SoCs in terms of area and power dissipation. Furthermore, high stability margins make the proposed design more attractive in nano-regime.

Chapter 5
SRAM Bitcell Design Using Unidirectional Devices

5.1 Introduction

Continued miniaturization or CMOS technology scaling has resulted an unprecedented increase in performance of single-core and multi-core microprocessors in the past four decades. However, the exponentially rising the transistor count has also increased the overall power consumption making performance per watt of energy consumption the key figure-of-merit for today's high-performance microprocessors and system-on-chip (SoC) products. Today, energy efficiency (performance per watt) serves as the central tenet of high performance microprocessor technology at the system and architecture level as well as the transistor level ushering in the era of energy efficient nano-electronics. Aggressive supply voltage scaling while maintaining the transistor performance is a direct approach towards reducing the energy consumption since it reduces the dynamic power quadratically and the leakage power linearly. In MOSFETs, the OFF-state leakage current (I_{OFF}) increases exponentially with reduction of threshold voltage. There are various leakage current mechanisms, such as band to band tunnelling (BTBT) at the drain-channel junction, the gate tunnelling leakage current through the ultra-thin gate dielectric and even direct tunnelling from source to drain are increasing with the continued scaling. Hence, there is a fundamental limit to the scaling of the MOSFET threshold voltage and hence the supply voltage. Scaling supply voltage limits the gate drive current (I_{ON}) and the I_{ON} to I_{OFF} ratio. The theoretical limitation to threshold voltage scaling mainly arises from MOSFETs 60 mV per decade subthreshold swing at room temperature and it significantly restricts low voltage operation.

It seems that quantum transistors such as Inter-Band *Tunnel Field Effect Transistors* (TFETs) or *HEterojunction Tunneling Transistors* (HETTs) may be a promising candidate to replace the traditional MOSFETs because the quantum

J. Singh et al., *Robust SRAM Designs and Analysis*, DOI 10.1007/978-1-4614-0818-5_5,
© Springer Science+Business Media New York 2013

tunnelling transistor has smaller dimension and steep subthreshold slope. Compared to MOSFET, TFETs have several advantages:

- Ultra-low leakage current due to the higher barrier of the reverse p-i-n junction.
- The subthreshold swing is not limited by 60 mV per decade at room temperature because of its distinct working principle.
- V_t roll-off is much smaller while scaling, since threshold voltage of TFET depends on the band bending in the small tunnel region, but not in the whole channel region.
- There is no punch-through effect because of reverse biased p-i-n structure.

One major difference between tunneling transistors (TFETs or HETTs) and traditional MOSFETs that should be considered in the design of circuits is uni-directionality. The tunneling transistors (TTs) exhibit the asymmetric behavior of current conductance. For instance, in MOSFETs the source and drain are interchangeable, with the distinction only determined by the biasing during the operation. While in TTs, the source and drain are determined at the time of fabrication, and the flow of current I_{ON} takes place only when $V_{DS} > 0$. For $V_{DS} < 0$ a substantially less amount of current flows, referred as I_{OFF} or leakage current. Hence, TTs can be thought to operate *uni-directionally*.

This uni-directionality or passing a logic value only in one direction has significant implication on realization of pass-transistors, transmission gates and SRAM designs. The pass-transistors and transmission gates require current to flow in both directions. These pass-transistors and transmission gates are also employed in SRAM designs as access devices, hence, possess significant restriction on implementation of SRAM design as well. However, asymmetric current flow does not posses any restrictions on the use of conventional static MOSFET logic circuit that employ pull-up network (PUN) and pull-down network (PDN). As such a logic circuit employs PUN and PDN in which current flow only in one direction either upward or downward, hence, they can be implemented without difficulty using unidirectional devices. Furthermore, the device characteristics as determined through the TCAD show an enhanced Miller capacitance as compared to MOSFETs can cause undesirable artefacts in the switching behavior of the TFETs.

Leakage power consumption in SRAMs has been a major concern in caches since ITRS projected that the percentage of memory in the SoCs will increase from the current 84% to as high as 94% by the year 2014 [48]. Low voltage operation is one of the most effective low power design techniques due to its quadratic dynamic and linear static energy savings. Lower threshold voltages increase the sub-threshold current exponentially and ultra thin gate oxides cause a huge increase in gate current. Various methods such as multiple threshold voltages and increased gate oxide thicknesses have been explored to reduce leakage in SRAMs. Adaptive or dynamic body biasing techniques have also been explored to reduce leakage power [101, 108, 122].

Recently, leakage reduction using steep subthreshold transistors has gained great attention. A steep subthreshold transistor allows us to operate at very low threshold voltages with ultra low leakage and low supply voltages (V_{DD}). Inter-band Tunnel

Transistors or 'TFETs' have been shown to be a promising steep subthreshold transistor which works on the principle of inter-band tunnelling [89]. TFETs have shown to be extremely power efficient in [59] for logic circuit applications. The authors in [59] also point out the problem of *uni-directionality* in TFETs and it's detrimental impact on 6T TFET SRAMs. Since uni-directionality has less impact on the logic circuit design as compared to SRAM bitcell design, where access or pass-gate transistors have to operate in both the directions to successfully read and write into the bitcell. To overcome this limitation, a 7T SRAM design was proposed in [59] with an extra read port to achieve higher stability margins. In this line, another novel 6T TFET SRAM bitcell was also proposed in [99] to overcome the problem of uni-directionality, and it achieves tolerable stability margins and performance at the same area of a CMOS conventional 6T CMOS SRAM design.

5.2 Tunneling Transistors

In the recent times, inter-band Tunnel Field Effect Transistors (TFETs) have been extensively investigated [14, 78, 89, 99, 110] due to their potential for sub-KT/q subthreshold slope device operation and thus enabling supply voltage reduction for low power logic applications. The Si/SiGe heterostructure uses gate-controlled modulation of band-to-band tunneling to obtain subthreshold slope of about 30 mV/decade with a large ON current. Figure 5.1 shows the optimized double gate device structure of a Si based N-channel and P-channel TFET proposed in [99]. An N-type TFET consists of a p+ source, intrinsic (i) channel and a n+ drain and the P-type TFET has n+ source, intrinsic channel and p+ drain regions. The source and drain regions are heavily doped regions with the channel region being intrinsic. The gate work function of N-channel TFET is modified suitably to obtain an equivalent P-channel TFET.

Figure 5.2 shows the band-diagram of a N-type TFET during the ON and OFF state. In the OFF state (i.e. $V_{GS} = 0$ V, $V_{DS} = 1$ V), the conduction in MOSFET is limited by the source side p-n junction barrier which prevents the thermionic

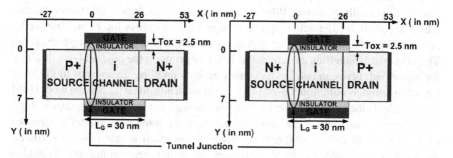

Fig. 5.1 An optimized structural model of double gate N-channel (NTFET) and P-channel (PTFE) Si-TFET proposed in [99]

Fig. 5.2 Band diagram
of a Si-NTFET under
ON and OFF conditions

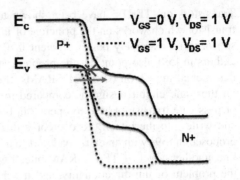

Table 5.1 Nominal values
of Si-TFET parameters
for optimized device
structures (NTFET
and PTFET)

Gate length, L_G	30 nm
Oxide thickness, T_{OX}	2.5 nm
Gate di-electric constant, ξ	21 (HfO_2)
Body thickness, T_{Si}	7 nm
Gate overlap	2 nm
Source/drain doping, $N_{S/D}$	10^{20} cm^{-3}
Channel doping, N_{Ch}	10^{15} cm^{-3}

emission of carriers. In the ON state (i.e. $V_{GS} = 1$ V, $V_{DS} = 1$ V), the source barrier is negligible enabling over the barrier thermionic emission. In contrast, TFETs operate by the tunnelling of carriers from the valence band in the source to the conduction band in the channel. In the OFF state (i.e. $V_{GS} = 0$ V, $V_{DS} = 1$ V), the transmission probability is low due to the thick depletion region associated with the source to channel tunnel junction resulting in very low OFF currents. With the application of the gate voltage (i.e. $V_{GS} = 1$ V, $V_{DS} = 1$ V), the depletion region shrinks and the carriers tunnel through the barrier. Since the TFET ON current is limited by the inter-band quantum mechanical tunnelling compared to thermionic emission over the barrier the ON current in silicon TFETs is much lower than MOSFETs. The reverse biased leakage current under the condition of OFF state (i.e. $V_{GS} = 0$ V, $V_{DS} = 1$ V) yields extremely low OFF current in the order of pico-femto amperes.

Table 5.1 shows the nominal parameters of optimized device structures. A non-local tunnelling model [96] is used for the simulation of tunnel current which accounts for the actual spatial charge transfer across the tunnel barrier by considering the actual potential profile along the entire path connected by tunnelling. The inter-band tunnelling current in the TFET depends on the potential profile along the entire path between two points connected by tunnelling. In contrast to the local tunnelling models commonly used [45,92], we use a non-local tunnelling model [46] which reflects the real space carrier transport through the barrier taking into account the potential profile along the entire tunnelling path. Band edge tunnelling masses of $m_c = 0.5 * m_0$ and $m_v = 0.65 * m_0$ (where m_0 is electron rest mass) for silicon are used to calculate the local imaginary wave numbers within the forbidden gap. Kane's two band model is then used to calculate the tunnelling probability.

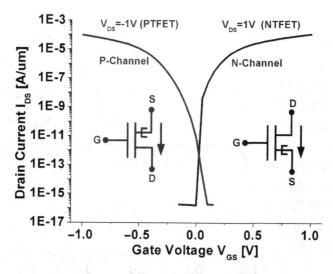

Fig. 5.3 $I_D - V_G$ characteristics of N-channel (NTFET) and P-channel (PTFET) and their symbols

The results presented here are obtained through drift-diffusion simulation where the Poisson and carrier continuity equations are solved self consistently. The inter-band tunnelling component is added to the carrier continuity equation as a generation-recombination (G-R) term. The G-R term contains adjustable scaling factors g_c and g_v kept at value equal to 0.1 and 0.4 respectively for Si which set the effective Richardson constant. We also obtained an excellent fit of our non-local tunnelling model with the experimental data from Fair and Wivell [32] for a reverse biased Si zener diode.

Figure 5.3 shows the $I_D - V_G$ characteristics of a Si NTFET and PTFET for $V_{DS} = 1$ V. For NTFET, we have obtained a $I_{DSAT} = 120 \,\mu\text{A}/\mu\text{m}$ and corresponding PTFET characteristics are also matched to it for the same drive current. The reverse biased leakage can be set to the order of pico-femto amperes by modifying the gate work function. We assume that the gate leakage is negligible due to the use of high-k dielectrics. We have also denoted the symbols for N-channel (N-TFET) and P-channel (P-TFET) in Fig. 5.3. The source side tunnelling barriers are represented by a bracket symbol and the current directions are also shown. In other words, current exiting from the source terminal is referred as an N-TFET, while current entering at the source terminal is referred as a P-TFET.

Figure 5.4 shows the $I_D - V_D$ characteristics of the same optimized device. The device exhibits expected characteristics due to tunnelling during positive V_{DS} (reverse bias conditions) while I_{DS} increases significantly for two conditions when V_{DS} is negative (forward bias). When V_{DS} is ~ -1 V, there is a significant I_{DS} irrespective of the value of V_{GS}. Significant current conduction is also observed when V_{DS} is slightly negative and V_{GS} is positive. This is due to electrons tunnelling from the conduction band of intrinsic 'i' region to the valence band of p+ source region.

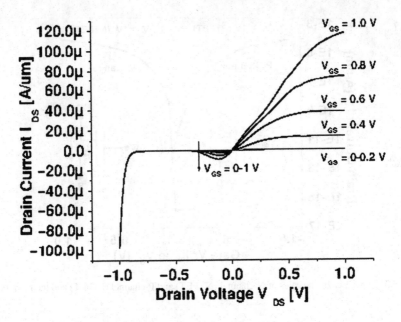

Fig. 5.4 TCAD simulated $I_D - V_D$ characteristics of a Si-NTFET

5.3   Development of TFETs Behavioural Model

Since analytical models for TFETs are not available and device simulation using Technology Computer Aided Design (TCAD) tools is not computationally efficient, which makes circuit level study almost impractical. Therefore, a look-up table based model using Verilog-A for circuit simulation and evaluation was developed. Also it makes simulation easy to parametrize some of the design parameters for exhaustive experiments. The Verilog-A module is then used as an instances for circuit simulation in Cadence Spectre. This efficient and accurate way of modeling is well suited for the emerging devices for which compact or SPICE models are not available [72]. In this model, I-V and C-V characteristics of the TFET devices were extracted through Sentaurus [96] TCAD simulation and stored as a two dimension look-up tables.

In order to demonstrate the effectiveness and accuracy of the developed behavioral (Verilog-A) model, DC and transient characteristics of the TFET devices and circuits (inverter) were simulated and compared with the TCAD results. The Si-NTFET device $I_D - V_D$ characteristics simulated in TCAD along with the Verilog-A model based simulated $I_D - V_D$ characteristics are shown in Fig. 5.5. It can be seen that the Verilog-A model is accurately captures the $I_D - V_D$ characteristics of a Si-NTFET device. At circuit level, DC and transient behaviors of a TFET inverter are evaluated for illustrating the effectiveness and accuracy of the Verilog-A model. Since characterization of inverter has direct impact on the functionality of a SRAM

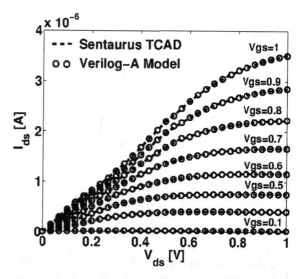

Fig. 5.5 Comparison of $I_D - V_D$ characteristics of a Si-NTFET obtained from TCAD and Verilog-A Model

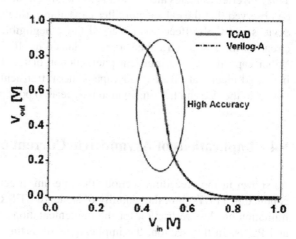

Fig. 5.6 DC voltage transfer characteristic (VTC) of a TFET inverter obtained from TCAD and Verilog-A model

bitcell design as it forms the basic unit of storing information and it can characterize its ability (or its basic functions) to read, write and hold the one bit of information. The TCAD and Verilog-A model based DC voltage transfer characteristics (VTC) of a inverter is shown in Fig. 5.6. It is clear that the Verilog-A model is accurately representing the TCAD based VTC.

As TFET devices exhibit the higher gate to drain Miller capacitance, C_{GD}, which leads to significant voltage overshoot and undershoot in its transient response and explain its physical origin from the energy band diagrams. These overshoots and undershoots have significant impact on the inverter delay, as a result SRAM bitcell performance may degrade. Therefore, transient input–output characteristics of a TFET inverter are simulated in TCAD and verified with the behavioral Verilog-A

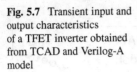

Fig. 5.7 Transient input and output characteristics of a TFET inverter obtained from TCAD and Verilog-A model

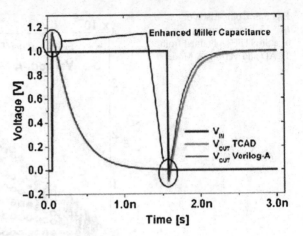

model, as shown in Fig. 5.7. An accurate modeling of the transient behavior of a TFET inverter validates the Verilog-A model for circuit level simulations. It can also be observed that the enhanced miller capacitance (high C_{GD}) values for the TFET devices and their effect was observed to be negligible (marked with circles) for circuits with high electrical effort as explained in [102] and [79]. Since, enhanced Miller capacitance is a transient phenomenon of the TFET devices and with the increased electrical effort (load capacitance), transient spikes goes down. But it increases the rise and fall time which may lead to poor performance of a circuit.

5.4 Implications of Asymmetric Current on SRAM Design

As shown in the preceding sections that the uni-directional (asymmetric) current conducting TFETs limit the viability of standard 6T SRAM bitcells. However, this limitation has less restriction for the implementation of logic circuits using PUNs and PDNs. In this section, the implications of asymmetric behavior of TFET on SRAM bitcell and its viability is explored to find the alternative solutions. In order to illustrate these implications of asymmetric behavior of TFET in SRAM bitcells, understanding of 6T CMOS SRAM bitcell current flow paths in different transistors under read and write operations are needed to be explained. Figure 5.8 shows the standard 6T CMOS SRAM bitcell storing '1' (i.e. node Q at V_{DD} and node QB at V_{SS}). To read the stored bit value, both the bit-lines (BL and BLB) are pre-charged to V_{DD} and then disconnected from the supply voltage followed by the word-line (WL) activation to high.

The read current path, as shown in Fig. 5.8a consists of transistor M_2 and M_6, pulls down the pre-charged bit-line (BLB), however, the node QB is maintained at ground by the transistor M_6. While the bit-line BL remains at V_{DD}, as node Q is held at V_{DD} by the transistor M_3. Therefore, difference of these bit-lines is sensed by the sense amplifier to determine the stored value. An important observation is that

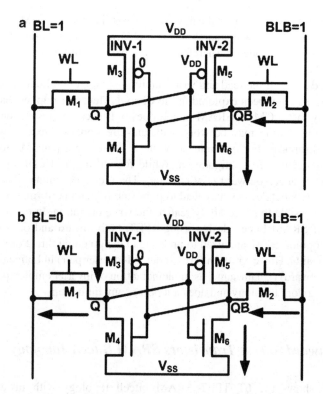

Fig. 5.8 Standard 6T CMOS SRAM bitcell current flow paths under read and write operations. (**a**) Under read operation. (**b**) Under write operation

under read operation there is no implication of uni-directionality, because only one current path exists at a time in one direction (i.e. either M_2 and M_6 or M_1 and M_4 have to conduct) which is in the inward direction. Hence, inward direction devices are good candidate for read operation. The read-stability (and speed) is determined by the sizing ratio of these transistors (M_6 to M_2 or M_4 to M_1), which is commonly referred as a cell ratio (β).

Similarly, for writing a value '0' in standard 6T CMOS SRAM bitcell the current paths are shown in Fig. 5.8b. The write operation or flipping the bitcell content from initially stored information (i.e. node Q at V_{DD} and node QB at V_{SS}) requires following actions. At the onset of write operation both the bit-lines (BL and BLB) are pre-charged followed by the word-line (WL) activation to high. In order to flip the bitcell content, the bit-line (BL) is driven to the ground potential and BLB is held at V_{DD} by a write driver. Node Q which is at V_{DD} will be pull-down to low through M_1 pass-gate transistor, while BLB which is held at V_{DD} will pull-up the node QB through M_2. In other words, access transistor M_2 aids in writing by pulling-up the node QB and access transistor M_1 helps in pulling down the node Q to low which makes SRAM bitcell flip more easily. It should be noted that for writing operation two current paths exist, one is in the inward direction (i.e. formed by M_2 and M_6) for pulling up the node QB, while another is in the outward direction (i.e. M_3 and M_1) for pulling down the node Q. Therefore, write operation through a uni-directional device may be difficult.

A detailed study of the read and write stability and their conflicting requirement of device sizing have been explored in the Chap. 1. Hence, it implies that the uni-directionality in 6T CMOS SRAM bitcell does not place any restrictions on read stability, while writing may be worst with uni-directional devices as one of the paths may disappear. For example, to write a '0', current path M_1 to BL will disappear, under inward configuration. While for writing a '1', current path M_2 will disappear, under outward configuration. Therefore, designing SRAM bitcell using uni-directional devices may lead to poor stability and performance or even it may not be possible to successfully realize the write operation. In order to see the feasibility of SRAM bitcell design using TFET, both inward and outward access transistors topologies are analyzed to understand the potential threat or viability of uni-directionality. However, latch (cross-coupled inverter pair) in both cases (either inward or outward) remain same, as implementation of inverter employ pull-up and pull-down network topology and does not posses any limitations.

5.4.1 Inward Access Transistors SRAM Bitcell Topology

Figure 5.9 shows the 6T TFET SRAM bitcell topology with inward access transistors, let's, assume bitcell storing a bit value '1'. In order to understand its functionality, both read and write operations are examined. To read a stored value from this bitcell topology, both bit-lines are pre-charged to V_{DD} and bit-line (BLB) is pulled down through the current flow path formed by M_2 and M_6, as shown in Fig. 5.9a. However, the bit-line, BL, remains at '1' due to absence of the discharging path and the difference of these bit-lines is sensed by the sense amplifier to determine the stored value. Therefore, 6T TFET SRAM bitcell based on inward access transistors topology will work satisfactorily. Figure 5.10 shows an example of Read Noise Margin (RNM) measurement obtained from the butterfly curve [94] for two different cell ratio (CR). When CR is 2, it indicates the successful read operation or adequate read noise margin (RNM), while, for CR is 0.2 it shows the read failure with just one stable point.

However, to write '0' (or flip the bitcell content) to this bitcell topology, access transistor M_1 cannot pull down the node Q, since, it conducts only in inward direction. As a result one conducting path (M_3 to M_1) is missing in this case. Therefore, access transistor M_2 must pull up the node QB without any assistance from the access transistor, M_1, (as it exists in standard 6T CMOS bitcell) to well above the trip-point voltage of the inverter (INV-1), as shown in Fig. 5.9b. Hence, write operation is performed by the single access transistor may substantially worsen the write-ability or in other words a significant amount of reduction in Write Noise Margin (WNM) can be experienced by the 6T TFET bitcell as compared to

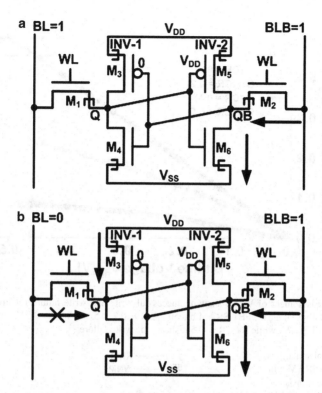

Fig. 5.9 6T TFET SRAM bitcell current flow paths for inward access transistors topology under read and write operations. (**a**) Under read operation. (**b**) Under write operation

Fig. 5.10 Measurement of RNM showing successful read and read failure for inward access transistor topology of a 6T TFET SRAM at CR = 0.2, 2 and $V_{DD} = 0.5$ V. A successful read is observed for CR = 2, while there is a read failure for CR = 0.2

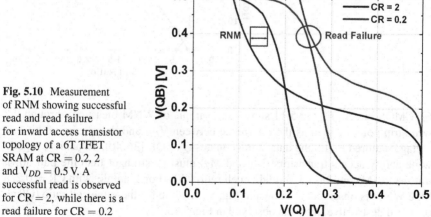

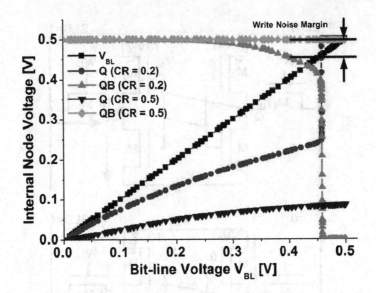

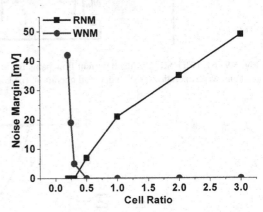

Fig. 5.11 Measurement of WNM showing successful write and write failure for inward access transistor topology of a 6T TFET SRAM bitcell at CR = 0.2, 0.5 and V_{DD} = 0.5 V. For CR = 0.2, there is a small WNM, while for CR = 0.5 it shows the write failure

Fig. 5.12 Noise margins for 6T TFET SRAM with inward access transistors topology at V_{DD} = 0.5 V

6T CMOS bitcell. Figure 5.11 shows an example of WNM measured through the write-trip point defined as the difference between V_{DD} and the maximum bit-line voltage required to flip the data storage nodes Q or QB [36, 40]. To achieve a good write-ability access transistors (M_1 and M_2) must be stronger than the pull down transistors (M_4 and M_6), i.e. smaller cell ratio (CR). For example, with CR = 0.2, a little WNM is observed, as shown in Fig. 5.11, while for the same cell ratio (CR = 0.2) read destructive failure is observed in Fig. 5.10.

Figure 5.12 shows the read and write noise margins (RNM and WNM) for different cell ratios of a 6T TFET SRAM bitcell topology with inward access transistors at V_{DD} = 0.5 V. The cell ratio is varied from 0.3 to 3 and both RNM

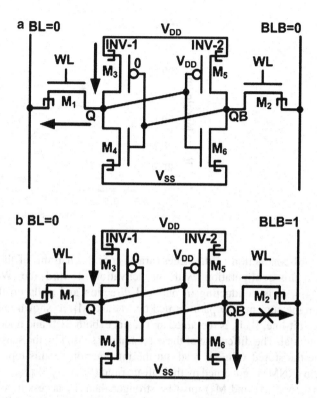

Fig. 5.13 6T TFET SRAM bitcell current flow paths for outward access transistors topology. (**a**) Under read operation. (**b**) Under write operation

and WNM were measured. It can be seen from Fig. 5.12 that the WNM reduces to 0 for cell ratio ($\beta = W_{Pull-Down}/W_{Access}$) > 0.3 while RNM is 0 for β < 0.3. Therefore, a 6T TFET SRAM bitcell topology with inward access transistors has enough read margin for cell ratio greater than 0.3, as it does not posses any restriction on read operation. But for the same variations in the cell ratio it has very poor WNM. Therefore, inadequate RNM and poor WNM for a practical cell ratio makes this topology impractical.

5.4.2 Outward Access Transistors SRAM Bitcell Topology

Figure 5.13 shows the 6T TFET SRAM bitcell topology with outward access transistors initially storing a bit value '1'. To read a stored value from this bitcell topology, both the bitlines are initially dis-charged to '0'. It is opposite in contrast with inward access transistor topology (or conventional 6T CMOS SRAM bitcell), where both bitlines are initially pre-charged to V_{DD}. Due to limitation imposed

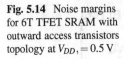

Fig. 5.14 Noise margins
for 6T TFET SRAM with
outward access transistors
topology at $V_{DD}, = 0.5$ V

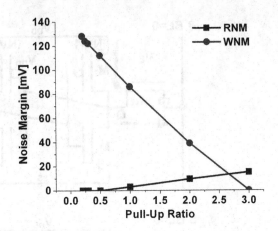

by the outward conduction of the access transistors discharging of the bitline is
considered in this configuration. With the activation of word-line (WL), access
transistor (M_1) starts conducting in outward direction and pulls up the bit-line
(BL) through the current flow path formed by M_3 and M_1, as shown in Fig. 5.13a.
However, the bit-line, BLB, is remained at '0', since, both BLB and node QB are at
the same potential. The difference of these bit-lines is sensed by the sense amplifier
to determine the stored value or read-out the information. In this topology, Read
Noise Margin (RNM) is governed by the pull-up ratio ($W_{Pull-Up}/W_{Access}$). Therefore,
pull-up transistors (M_3 and M_5) must be stronger than the access transistors (M_1
and M_2), in order to prevent the read failure. Otherwise access transistors (M_1)
may easily drain out the node Q to ground potential. Similarly, when node QB at '1'
access transistor M_2 may discharge the node QB to ground potential via bitline BLB.

However, to write into this bitcell topology, both the bit-lines are initially
precharged to V_{DD} and during the write cycle one of the bit-line has to be driven to
ground potential by the write driver. In order to write '0' to this bitcell configuration,
that is, node Q is pulled down well below the trip-point of inverter 2 (INV-2)
by driving BL to ground potential by outward conducting access transistor M_1.
However, bitline BLB is remains at '1', since, outward access transistor M_2 cannot
conduct in inward direction to pull-up the node QB, as shown in Fig. 5.13b.
Therefore, write operation in this topology is un-aided, or in other words, node QB
is not pulled-up by the bitline BLB due to non-conduction of access transistor M_2.
Furthermore, in this configuration, Write Noise Margin (WNM) is governed by the
pull-up ratio ($W_{Pull-Up}/W_{Access}$). For a successful write operation, access transistors
(M_1 and M_2) must be stronger than the pull-up transistors (M_3 and M_5). It clear that
the write operation is performed by one access transistor only, and not differentially
aided by the another access transistor.

Figure 5.14 shows the RNM and WNM of a outward access transistor 6T TFET
SRAM bitcell topology for different pull-up ratios at $V_{DD} = 0.5$ V. It can be observed
that the RNM starts increasing only for pull-up ratios ($W_{Pull-Up}/W_{Access}$) greater
than 1.5 while WNM reduces rapidly with increase in the pull-up ratio. It leads

to conflicting sizing requirements for read and write operations. In other words, for a successful read operation M_3 must be stronger as compared to M_1, while for a write operation M_1 must be stronger than the M_3. Similarly, when node QB is at '0' a successful read operation require stronger M_5 as compared to M_2 and for a write operation stronger M_2 is required as compared to M_5. An important observation is that both read and write operations have conflicting sizing requirements. For example, a success full read operation requires a stronger M_3 as compared to M_1, while for a write operation M_1 must be stronger than the M_3.

Thus, a 6T TFET SRAM bitcell with either inward or outward access transistor configuration is not possible due to poor RNM and WNM for a range of cell ratio and pull-up ratio. Furthermore, both the inward and outward access transistor topologies have conflicting device sizing requirements, which is further worsened by the un-aided read or write operation. A 7T TFET SRAM with outward access transistor configurations was proposed in [102]. In this design, outward access transistor configuration is used to obtain the adequate write margin while the read margin is improved by providing a read-buffer with an extra transistor and separate read bit-line and word-line. In this line, a case study of 6T TFET SRAM cell is presented in the next section.

5.5   A Case Study of a 6T TFET SRAM Bitcell Design

As shown in the previous section that the practical 6T TFET SRAM bitcell design is not feasible, due to poor RNM and WNM constraints imposed by the uni-directionality and the conflicting devices sizing requirement in both (inward and outward access transistor) topologies. In order to address these issues, recently two SRAM bitcells were proposed [99, 102]. In this section, a case study on 6T TFET SRAM bitcell is presented [99], this design has minimum number of devices and preserves the adequate RNM and WNM. The 6T TFET bitcell design is consists of cross coupled inverters (INV-1 and INV-2) with the bit-lines BL and BLB connected to node Q through the access transistors M_1 and M_2 (Note that both access transistors are connected to the same node Q), as shown in Fig. 5.15. It is a design strategy to connect both bit-lines to node Q for writing either '1' or '0', in order to provide the virtual ground (VV_{SS}) to INV-1 to assist the write operation. This virtual grounding helps in improving the WNM (i.e. write-ability), by decoupling (or weakening of the re-generative action) of the cross-coupled inverters.

The TFET 6T SRAM bitcell employ combination of both (inward and outward access transistor) topologies for adequate RNM and WNM, as shown in Fig. 5.15. It is shown in the previous section that the inward access transistor topology yields better RNM, so that M_1 is connected in inward direction to provide better RNM. However, outward access transistor topology provides better WNM, therefore, M_2 is connected in outward direction. Employing both inward and outward access transistors helps in realizing the 6T TFET SRAM bitcell.

Fig. 5.15 The 6T TFET
SRAM bitcell design with
inward and outward access
transistors connected
to node Q

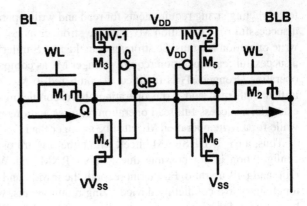

Fig. 5.16 Read current path
of a 6T TFET SRAM bitcell
design

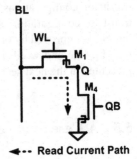

5.5.1   Read Operation in 6T TFET SRAM Bitcell

A differential read operation is presented in this design to achieve better read
performance. The read cycle starts with pre-charging of both bit-lines (BL and
BLB) to V_{DD} and then the WL is asserted to high. Initially, it is assumed that
the bitcell stores '0' at node Q. In this case, inward access transistor (M_1) starts
conducting in inward direction and pulls down the bit-line (BL) through the current
flow path formed by M_1 and M_4 transistors. However, BLB remains at V_{DD} and it
will not pull-up the node Q since, outward transistor M_2 will not conduct in inward
direction due to drain and source polarities. The difference between BL and BLB
is sensed by the sense amplifier to read-out the stored data. Figure 5.16 shows the
current path during a read operation formed by the access transistor (M_1) and pull-
down transistor (M_4) in the 6T TFET SRAM bitcell design. Therefore, Read Noise
Margin (RNM) in this design is governed by the ratio of W_{M_4}/W_{M_1}, and for a better
RNM, M_4 must be stronger than the M_1. Similarly, when data storage at node Q
is '1', bitlines BL and BLB remain at V_{DD} and appropriate information can be
distinguished by the sense amplifier.

The inward access transistor (M_1) for read operation in the proposed bitcell
design is chosen, since this configuration allows a higher RNM than the outward

Fig. 5.17 Write operation equivalent path of a 6T TFET SRAM bitcell for (a) write '1' and (b) write '0'

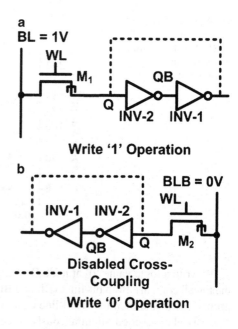

access transistor configuration, as shown in Sects. 5.4.1 and 5.4.2 (Figs. 5.12 and 5.14). While WNM is significantly improved by providing the virtual ground, explained in the later sections.

5.5.2   Write Operation in 6T TFET SRAM Bitcell

Write operation in the 6T TFET bitcell design is done through either one of the access transistors depend on input data. To write a '1' onto node Q (initially Q at '0'), both the bit-lines (BL and BLB) are charged to V_{DD} and then followed by the word-line enable signal, WL = '1'. The write control logic also enables the virtual ground simultaneously. This weakens the inverter, INV-1, and disables the cross-coupling between the two inverters INV-1 and INV-2. The inward access transistor M_1 pulls the node Q to high via bitline BL. Once node Q reaches to the trip-point of the INV-2, QB becomes '0', and VV_{SS} line is connected to ground potential so that the cross-coupled inverters start working in order to maintain the written information. Figure 5.17a shows the equivalent circuit diagram comprises of series connected inverters (cross coupling is disabled by the virtual ground as shown by the dotted line) and an inward access transistor, M_1, during write '1' operation. Hence, write '1' operation is performed via access transistor, M_1, while access transistor, M_2, does not works due to its outward direction configuration and not shown in the equivalent diagram. Also pre-charging of bit-line, BLB, is essential so that node Q should not be drained out simultaneously during the write '1' operation, by the access transistor M_2.

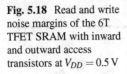

Fig. 5.18 Read and write noise margins of the 6T TFET SRAM with inward and outward access transistors at $V_{DD} = 0.5\,V$

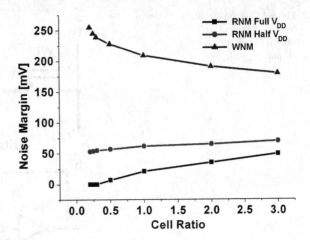

To write a logic value '0' at node Q, when it is initially at logic '1' (i.e. flipping the bitcell content). Both bit-lines (BL and BLB) are dis-charged to ground potential ground (V_{SS}) and then the word-line enable signal, WL = '1', and virtual ground VV_{SS} is also asserted simultaneously to disable the cross-coupling between the two inverters INV-1 and INV-2. The write '0' operation is performed via access transistor, M_2, while access transistor, M_1, does not works due to its inward direction configuration. Also dis-charging of bit-line, BL, is essential so that node Q should not be pulled up simultaneously during the write '0' operation. Figure 5.17b shows the equivalent circuit diagram along with disabled cross-coupling mechanism shown in dotted during write '0' operation.

In order to demonstrate a successful read and write operation, the RNM and WNM of the 6T TFET SRAM bitcell are simulated in HSPICE for different cell ratios, CRs (β) at $V_{DD} = 0.5\,V$, while, the pull up ratio is kept at minimum. In Fig. 5.18, the RNM at half pre-charged and fully pre-charged bit-line is also measured, it can been that the half pre-charged bit-line yields much better RNM than fully pre-charged bit-line. For CR, $\beta > 2$, there is no significant improvement in the RNM while a slight degradation in the WNM is observed, also making bitcells with higher cell ratios will increase the bitcell area. Hence, all the simulation results presented in the next section uses cell ratio (β) of 2 unless specified. Due to the asymmetric nature of the design, writing a '1' is more difficult than writing '0', hence, only measurement of WNM for writing a '1' into the bitcell is considered in this case study.

5.6   SRAM Bitcell Design Metrics

Stability, performance, power and area are the key bitcell design metrics widely used in the silicon industry to identify a potential SRAM bitcell design, particularly in the nanometer regime. For comparison, standard 6T CMOS SRAM and 7T

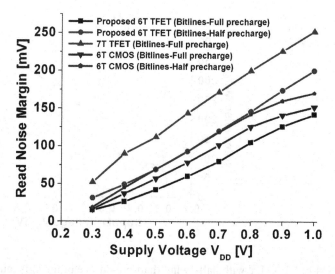

Fig. 5.19 A comparison of read static noise margins (SNMs) for different SRAM bitcell designs at various supply voltages V_{DD}

TFET SRAM bitcell designs are used. The 32 nm Predictive Technology Models (PTM) [88, 123] for 6T CMOS, while the 6T TFET and 7T TFET [102] SRAM bitcells are simulated with the same device as explained in Table 5.1. In this section, different bitcell design metrics are evaluated and compared.

5.6.1 SRAM Bitcell Stability

An adequate read stability and write-ability of a SRAM bitcell are highly desirable for successful realization of robust and high performance caches. The RNM and WNM are the widely used metrics for stability analysis of a SRAM bitcell. Figure 5.19 shows the RNM of different bitcell designs. For the 6T TFET and 6T CMOS bitcells both the bit-lines, BL and BLB, are pre-charged to full V_{DD} and half V_{DD} to evaluate the RNM. While, for 7T TFET bitcell, pre-charging of bitlines to any level does not have any impact on the RNM due to its isolated read-port or a separate read word-line. Therefore, 7T TFET SRAM bitcell shows the highest RNM, because of an isolated read-buffer which yields the RNM equivalent to Hold Static Noise Margin (SNM). The isolated read buffer concept has been widely explored in CMOS SRAM designs to improve the RNMs at the cost of silicon overhead, and referred as read noise margin free SRAM bitcells. However, the 6T TFET with fully pre-charged bit-line has the lowest RNM. This is because of the single access transistor which conducts during the read operation and rises the internal node (Q) voltage to a higher value than a 6T CMOS SRAM, while the other access transistor does not assists because of its uni-directionality.

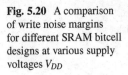

Fig. 5.20 A comparison of write noise margins for different SRAM bitcell designs at various supply voltages V_{DD}

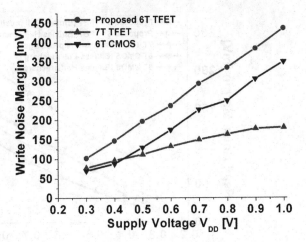

The RNM of 6T TFET with half-swing (half pre-charge bitlines) is much better than the 6T CMOS with half and full pre-charged bit-lines. In 6T CMOS SRAM, half pre-charged bit-lines are not as effective as 6T TFET SRAM. This is due to the symmetric nature of SRAM where one of the bit-lines connected to a node (Q or QB) via access devices storing a V_{DD} is also pre-charged to half V_{DD}. Hence, this scenario is not effective in holding that node at V_{DD} as compared to pre-charging to full V_{DD} due to conduction from the node to bit-line in the former case. However in 6T TFET design, M_2 in Fig. 5.15 does not conduct in the reverse direction and this contributes to higher RNM at half pre-charged V_{DD}. At $V_{DD} = 0.3$ V, a 63% improvement in RNM is observed over a 6T CMOS while it is 59% less than the 7T TFET bitcell. The advantage of higher RNM in 7T TFET bitcell is purely from the extra transistor used as a read port.

Figure 5.20 shows the WNM of SRAM bitcell designs for different V_{DD}. The WNM of the 6T TFET SRAM design is higher than its counterpart designs due to virtual ground mechanism used for weakening of the cross-couple inverters which enables a faster write operation. At $V_{DD} = 0.3$ V, a 46% and 32% improvement in WNM over 6T CMOS and 7T TFET bitcells, respectively, is observed.

5.6.2 SRAM Bitcell Performance

Read and write delays are the metrics used to compare the performance of different SRAM bitcell designs. In 6T CMOS and 6T TFET bitcells, read delay is defined as the time delay between 50% of word-line (WL) activation to 10% of pre-charged voltage difference between the bit-lines. In 7T and 8T SRAM bitcell designs, bit-line sensing is done using CMOS logic gates and not by using differential sense amps [28, 77]. So, for the 7T TFET bitcell, read delay is measured between 50% of word-line (WL) activation to 50% of pre-charged bit-line voltage. Figure 5.21 shows

Fig. 5.21 A comparison
of read delay for different
SRAM bitcell designs
at various supply voltages
V_{DD}

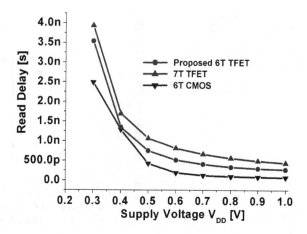

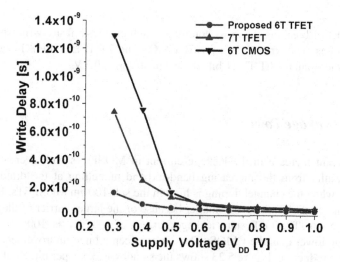

Fig. 5.22 A comparison of write delay for different SRAM bitcell designs at various supply
voltages V_{DD}

the read delay of different SRAM bitcell designs. It is observed that CMOS performs
better than TFETs in the entire voltage range due to it's high drive current. At
$V_{DD} = 0.3$ V, 6T CMOS design has a better read delay than 6T TFET and 7T TFET
by 40% and 58% respectively. However, this problem can be solved in TFETs by
moving to lower band-gap and low effective mass materials such as Indium Arsenide
(InAs) which have a higher tunnelling rate through the barrier and higher drive
current (I_{ON}) of ~85 μA/μm for $V_{DD} = 0.25$ V [78].

The write delay is defined as the time between the 50% activation of the wordline (WL) to when the internal Q is flipped to 90% of its full swing. At lower
voltages, write delay of the 6T TFET SRAM design is significantly less than the
6T CMOS and 7T TFET SRAM designs as shown in Fig. 5.22. This is due to the

Fig. 5.23 Standby leakage power of 6T CMOS and 6T/7T TFET SRAM bitcell designs for different supply voltages

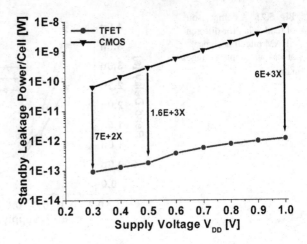

simple fact of breaking the cross coupling which enables a faster write speed than other designs. The write delays for 6T CMOS and 7T TFETs are 8.1× and 4.7× times higher than the 6T TFET bitcell design at $V_{DD} = 0.3$ V.

5.6.3 Leakage Power

A significant increase in the leakage current in MOSFETs with decreasing gate length results from the increasing band-to-band tunnelling at the drain-channel junction, when the channel doping is high in the sub-100 nm MOSFETs. However, the leakage in TFETs is much smaller because of the larger barrier of the reversed p-i-n junction. The OFF state leakage current of a TFET is 100s of order of magnitude lower than the CMOS. Thus, it can seen a huge improvement in terms of leakage reduction. Figure 5.23 shows the standby leakage per bitcell of different SRAM bitcell designs. Both 6T and 7T TFET bitcells have equal leakage power due to the presence of the same leakage paths. Assuming that the node connected to the read-port held at '0'. A 700× and 1,600× improvement in leakage reduction over 6T CMOS bitcell at 0.3 V and 0.5 V V_{DD}, is achieved. This shows that TFETs are a potential replacement candidate for CMOS transistors at low voltage and low power applications.

5.6.4 Area

A comparison of the layout for the standard 6T CMOS, the proposed 6T TFET and 7T TFET bitcells is also shown; the layouts adhere to the "thin bitcell" topology [56], alleviating lithography stresses and device mismatch sources by

Fig. 5.24 Efficient SRAM bitcell layouts for (**a**) standard 6T CMOS bitcell, (**b**) the proposed 6T TFET bitcell and (**c**) 7T TFET bitcell allowing sharing of most abutting source and drain junctions and poly between adjacent bitcells

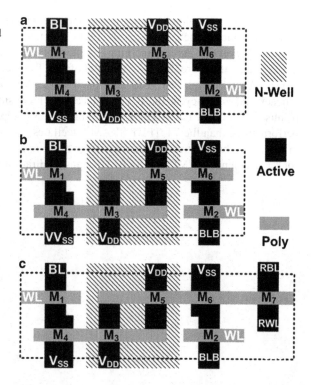

minimizing jogs in the poly. As shown in Fig. 5.24, these layouts are highly efficient in the sense that they amenably allow sharing of all source and drain junctions and poly wires with abutting cells. However, these baseline layouts can change dramatically to optimize for low-voltage operation and high data stability in the presence of process variation.

The 7T TFET SRAM bitcell is bound to have an increase of around 15% [102] area overhead over the standard 6T SRAM bitcell. In the isolated read-port 8T SRAM bitcell, two read-port transistors from the adjacent bitcells can be abutted in 7T TFET SRAM bitcell, making the overhead for two 7T bitcells equal to that of one 8T bitcell. While 8T SRAM bitcell exhibits 30% area overhead as compared to standard 6T bitcell [24]. However, the proposed 6T TFET SRAM bitcell can be realized without increase in silicon overhead, as shown in Fig. 5.24.

5.7 Summary

In Si-TFET, a higher potential barrier of the reverse p-i-n junction yields ultra-low leakage current, which makes Si-TFET as an ideal solution for low power SRAM design. In this chapter, therefore, a case study on 6T Si-TFET SRAM bitcell design to enable ultra-low voltage and low power applications is presented. The asymmetric current conduction behaviour of Si-TFETs that makes SRAM design difficult is addressed carefully. As an emerging device compact or analytical models are yet not available, therefore, behavioral model of TFET devices using industry standard Verilog-A was developed. The developed model was tested for computational efficacy and accuracy at the device and circuit level simulation. The simulation results show that the 6T Si-TFET SRAM bitcell has comparable margins and better performances than the 7T TFET SRAM bitcell design. A significant improvement in leakage reduction over the entire voltage range makes TFETs suitable candidate for replacement for CMOS in SRAM designs at ultra-low voltages such as 0.3 V.

Chapter 6
NBTI and Its Effect on SRAM

6.1 Introduction

In SRAM, random dopant fluctuations and critical dimension variations are not the only challenges that affect the reliability. Over the life time of a system, transistor I-V characteristics can be affected by phenomena such as Negative Bias Temperature Instability (NBTI), hot carrier effects, or time dependent dielectric breakdown. These changes to I-V characteristics from such phenomena affect the SRAM metrics and reliability. The NBTI is particularly a challenging problem because it not only changes the strength (reduction in trans-conductance and drain current) of PMOS transistor, but it does so over the life time of a system. It makes difficult to screen out SRAM bitcells that function immediately after processing but which will eventually fail due to NBTI. In nano-regime, this systematic reduction in PMOS transistor strength, due to NBTI over the life time severely degrades the SRAM metrics and poses a significant reliability concern as well as a limiting factor in future device scaling [48,60].

In particular, for sub-threshold operating devices that demand for higher drive current, NBTI significantly enhances the threshold voltage and thereby reduction in drive current simultaneously [93]. The NBTI-induced shift in threshold voltage degrades the performance of CMOS digital circuits. However, degradation in performance can be off-set by up-sizing of the PMOS devices during the design phase. Subsequently, this leads to an increase in power and silicon area overhead. The area, power and performance trade-offs due to NBTI effect have been widely studied in [50, 62, 74, 87, 115]. In SRAM bitcells, these trade-offs do not work effectively, because up-sizing of the devices increase the power budget and reduces the cache density. Furthermore, stability of the SRAM bitcell is becoming a more sensitive issue and drastically shifts the Static Noise Margin (SNM) and Write Noise Margin (WNM) over the time due to aging effects.

The impact of NBTI on SRAM bitcell reliability and stability is a subject of recent interest. In [90], authors showed that NBTI could affect read SNM by as much as 8% at $V_{DD} = 0.8$ V, and the effect is more dominant at lower V_{DD}. Authors in [65], extended this study and show that the variation in read stability increases as well, leading to larger failure counts in an SRAM array. These results demonstrate that NBTI has a significant impact on bitcell reliability. However, several researchers have reported a relatively low impact of NBTI on shift in threshold voltage of PMOS [17, 71, 82] at higher V_{DD}. Although such low sensitivities seem to contradict with results reported in [9, 64, 65, 90]. An important observation is that the low sensitivities were reported at relatively higher V_{DD} (≥ 0.9), and may have underestimated the NBTI impact at lower voltages.

SRAM bitcells are also particularly more susceptible to the NBTI effect because of their topologies. Since, one of the PMOS transistors is always negative biased if the bitcell contents are not flipped, it introduces asymmetry in the standard 6T SRAM bitcell due to a shift in threshold voltage in either of PMOS devices. In asymmetric SRAM bitcell, one of the Voltage Transfer Curves (VTCs) moves horizontally, as a result one of the butterfly curve lobe becomes smaller than the other which makes the read SNM poor, and more susceptible to process variation and NBTI induced failures.

The effect of NBTI independently and along with process variations on standard 6T (symmetric and asymmetric) SRAM bitcells and read static noise margin free 6T SRAM bitcell based cache configurations, are investigated in this chapter. In standard 6T SRAM bitcell, symmetric and asymmetric (dual-V_{TH}) SRAM bitcell based caches are considered. Symmetric 6T SRAM bitcells are generally used for implementation of high performance and high density caches. While, asymmetric SRAM bitcell based caches, designed with dual-V_{TH} technology have been recently studied for their strong potential of leakage power savings [8, 81]. In asymmetric SRAM bitcells, sub-threshold leakage current devices are made of high V_{TH}, while assuming the skewed distribution of storage bit '0'. The read static noise margin free SRAM bitcells consist of an isolated read-port and a cross-coupled inverter pair have recently attracted a lot of attention [25, 98, 108]. In these bitcells data storage nodes are isolated by providing a separate read-port (read current path), hence, there is no degradation in read SNM during the read cycle, referred as read SNM free SRAM bitcells.

It is investigated that employing different power saving strategies to the caches can recover a substantial portion of the stability noise margins (SNM and WNM) lost due to the predominant occurrence of logic '0' being stored in caches. Based on different power saving strategies proposed in [69], six different cache configurations are formed and their duty cycles are derived from the average inactive time experienced by the cache blocks for different applications. This leads to an additional design consideration while determining which of the power saving strategies should be applied in cache design, particularly if lifetime operation is a prime concern. Furthermore, study of the intra-die process variations employing different SRAM bitcells based caches for different cache configuration is also done along with the NBTI.

6.2   The Physics of Negative Bias Temperature Instability (NBTI) and Its Impact

The NBTI effect becomes a practical concern in nano-regime CMOS technologies due to exponential dependence on oxide thickness and temperature [27]. NBTI causes an absolute increase in PMOS devices threshold voltage which contributes to degradation of the mobility, transconductance and drain current. Although NMOS devices can be damaged due to NBTI stress, however, the damage is not activated in the operational configuration of the NMOS device, and hence, PMOS devices typically receive high emphasis. The NBTI phenomenon that causes the absolute increase in threshold voltage (V_{TH}) of a PMOS transistor due to the formation of interface traps with respect to time. Under the negative bias condition (i.e. $V_{GS} = -V_{DD}$) of a PMOS transistor, interface states and traps are generated as the hydrogen diffuses toward the gate. This phase of NBTI is called stress, as shown in Fig. 6.1a. The NBTI causes interface states and traps fixed positive charge in the oxide, the question naturally arises: "How do these interface states and fixed charges affect the device operation?"

From the physics of MOSFET device, the threshold voltage of a device is proportional to the number of charges over the capacitance of the gate oxide. The charges in a device can be fixed positive and that have populated an interface state. Changing the fixed charges and interface states within the device will result in a V_{TH} shift. As a result, other device parameters such as the drain current and transconductance are subsequently affected. Drain current degradation due to V_{TH} shifts impact matched devices and analog circuits greatly. It also degrade device performance in a circuit, as a result, leading to timing issues and circuit failure.

In Fig. 6.1b, when voltage of the gate is set to V_{DD} then no new interface traps are generated while hydrogen diffuses back and anneals the broken bonds. However, full recovery of the traps becomes impossible since hydrogen is no longer available and this is referred as a recovery phase. Thanks to annealing process for dynamically recovering the threshold voltage during the recovery phase and thereby a significant amount of performance and circuit stability can be recovered. A circuit designed for skewed activity, while keeping the dynamic recovery of threshold voltage in mind can greatly reduce the ageing effect due to NBTI. Specifically for SRAM bitcells where one of the PMOS transistors is always in stress mode.

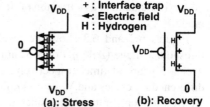

Fig. 6.1 PMOS biasing transistor in the stress and recovery phases due to NBTI

It is believed that the NBTI phenomenon involves a Reaction–Diffusion (R-D) process, and that the stress and recovery phases are successfully analyzed using the R-D model available on the PTM website [88, 124], and explained in more details in [3, 107], as follows:

$$\Delta|V_{TH}| = \sqrt{K_v^2(t-t_0)^{0.5} + \Delta V_{TH0}^2} + \delta_v \quad (stress)$$

$$= (\Delta V_{TH0} + \delta_v)\left[1 - \sqrt{\eta(t-t_0)/t}\right] \quad (recovery)$$

where

$$K_v = AT_{ox}\sqrt{C_{ox}(V_{GS} - V_{TH})}.exp\left(\frac{E_{ox}}{E_o}\right).$$

$$\left[1 - \frac{V_{DS}}{\alpha(V_{GS} - V_{TH})}\right].\left(-\frac{E_{ox}}{E_o}\right)$$

and

$$E_{ox} = (V_{GS} - V_{TH})/T_{ox} \qquad (6.1)$$

where, change of V_{TH0} (ΔV_{TH0}) is due to stress or recovery happens at $t = t_0$, the process and design parameters are V_{DS}, V_{GS}, V_{TH}, T and T_{OX} are scalable with model. The constants η, α, E_0, δ_v and E_a are extracted from the technology parameters.

According to Eqs. 6.1, there are a number of factors that influence the shift in threshold voltage due to NBTI. There is a positive relationship among these parameters such as temperature, supply voltage, the duty cycle and the magnitude of change in the threshold voltage. By lowering any of these parameters, reduction in change-in-threshold ($\Delta|V_{TH}|$) voltage will also be observed. Temperature is a function of power density and the rate at which heat is being removed from the system. Supply voltage can also be modified in order to control the power density because of its quadratic relationship, depending upon the workload for increased or decreased performance. Under lighter workloads the supply voltage can be lowered and conversely increased for demanding workloads.

Since, the PMOS only degrades when there is a negative gate to source (V_{GS}) voltage difference exist, then the ratio of negative V_{GS} to positive V_{GS} is referred as the duty cycle (β), it heavily impacts the change in threshold. For cache, this duty cycle is defined as a fraction of time cache is active as compared to idle (sleep) mode.

Figures 6.2 and 6.3 show the resulting (stress and recovery phases) shift in threshold voltage ($\Delta|V_{TH}|$) of a 32 nm and 45 nm PMOS transistor due to NBTI over a 5 years of time, respectively. The shifted threshold voltage is plotted for different duty cycles and $V_{GS} = -V_{DD} = -1$ V (-0.9 V for 32 nm) and T $= 125°$C. An elevated temperature is considered for simulation because NBTI has detrimental impact at higher temperature as compared to moderate one. Duty cycles for different

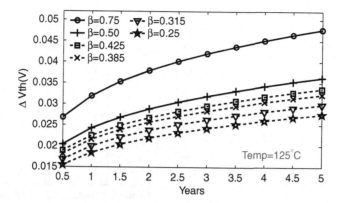

Fig. 6.2 Shifted threshold voltage ($\Delta |V_{TH}|$) due to NBTI for a 32 nm technology node PMOS transistor versus time for different duty cycles (β)

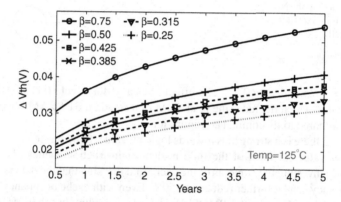

Fig. 6.3 Shifted threshold voltage ($\Delta |V_{TH}|$) due to NBTI for a 45 nm technology node PMOS transistor versus time for different duty cycles (β)

cache configurations are explained and derived in Sect. 6.5. It can be seen from Figs. 6.2 and 6.3 that the duty cycle (β) has significant role in modulating the threshold voltage of a PMOS transistor. Higher duty cycle has more impact on shifting the resulting threshold voltage as compare to smaller duty cycle. In 5 years of time span, shift in threshold voltage for duty cycle of $\beta = 0.25$ is 28 mV, however, for the same time period and duty cycle of $\beta = 0.75$, shift in threshold voltage is 47 mV. Hence, it could be a good candidate to control the aging effects and its associated impacts on circuits and systems.

Two extreme duty cycles ($\beta = 0.25$ and $\beta = 0.75$) are considered to see the impact of NBTI on read SNM of a standard 6T SRAM bitcell. The lower duty cycle ($\beta = 0.25$) corresponds to least NBTI impact, where SRAM is subjected to a minimum activity under the NBTI. However, higher duty cycle ($\beta = 0.75$) corresponds to heavy NBTI impact on the SRAM bitcell when logic bit value '0' being stored on average 75% of bit value. Figure 6.4 shows the NBTI impact (i.e. degradation of read SNM) on standard 6T SRAM bitcell for two different duty

Fig. 6.4 Standard 6T SRAM
bitcell read static noise
margin (SNM) degradation
due to NBTI for $\beta = 0.25$ and
$\beta = 0.75$

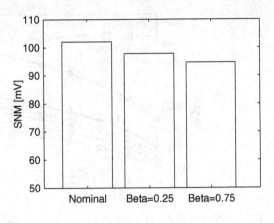

Fig. 6.5 Simulation setup
used for modelling the NBTI
effect in PMOS transistor of a
SRAM bitcell

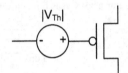

cycles $\beta = 0.25$ and $\beta = 0.75$. The effect of heavy duty cycle (i.e. $\beta = 0.75$) is clearly visible which degrades the read SNM of a standard 6T SRAM bitcell about 10% as compared to its nominal configuration.

In caches, there is a strong bias towards logic bit value '0' being stored on average 75% of bit value for most of the time in data or instruction caches [21, 81]. By periodically inverting the contents of the cache and marking that the data is inverted this occupancy can be further reduced to 50%. Even with cache occupancy of logic bit value '0' is 50%, then the PMOS devices are degrading but their degradation is occurring in a balanced fashion. Asymmetric SRAM bitcell based caches have been developed to take the advantage of skewed distribution of logic bit value '0' in caches for leakage power reduction. However, asymmetry introduced by the dual-V_{TH} devices seriously degrades the read SNM and make the SRAM bitcell more vulnerable to NBTI, process variations, soft errors and loss of stability, as shown in next section.

6.3 NBTI Model

To analysed the effect of NBTI on the degradation of SRAM read and write noise margins due to shift in threshold voltage of the PMOS transistors of a SRAM bitcell. The SRAM bitcells have been simulated considering the NBTI-induced shift in threshold voltage in each of the PMOS transistors. The NBTI degradation is modelled as a voltage source in series with PMOS gate [49, 91], as shown in Fig. 6.5. The shift in threshold voltage was calculated from the previously discussed R-D model for different duty cycles and time spans.

6.4 SRAM Bitcells Under NBTI

Figure 6.6 shows the symmetric standard 6T SARM bitcell and an asymmetric 6T SRAM bitcell for reduced leakage current is shown in Fig. 6.7. The aging phenomenon in different SRAM bitcells was introduced by incorporating the shift in threshold voltage of the PMOS devices, which was calculated from the R-D model discussed in the previous section and it was developed in MATLAB. Shift in threshold voltage was incorporated in different SRAM bitcells using the NBTI model as shown in Fig. 6.5. In symmetric 6T SRAM bitcell simulation setup, shown in Fig. 6.6, transistor M_1, M_2, M_4, M_5 and M_6 have nominal value of V_{TH} models, while shifted $\Delta|V_{TH}|$ value due to NBTI is used for M_3 transistor model. Since, V_{GS} of M_3 is $-V_{DD}$ as a result it will experience the NBTI effect at large as compared to M_5.

In asymmetric 6T SRAM bitcell, shown in Fig. 6.7 has three types of transistor models:

- Transistors M_1 and M_6 have nominal value of V_{TH} models,
- Shifted $\Delta|V_{TH}|$ value due to NBTI is used for M_3 transistor model, and
- Transistors M_2, M_4 and M_5 have high value of V_{TH} models to reduce the leakage current.

All transistors in symmetric and asymmetric 6T SRAM bitcells are of minimum feature sized with bitcell ratio $= 2$ and $M_1 = M_2 = M_3 = M_5 = 45\,\text{nm}/45\,\text{nm}$ (32 nm/32 nm), and $M_4 = M_6 = 90\,\text{nm}/45\,\text{nm}$ (64 nm/32 nm). In the read SNM free 6T (or 8T) SRAM bitcell, as shown in Fig. 6.9 regular-V_{TH} and minimum feature sized transistors are used for simulation, while shifted $\Delta|V_{TH}|$ value due to NBTI is incorporated with M_3 transistor model.

Figure 6.8 shows the Voltage Transfer Characteristics (VTCs) or butterfly curves of 6T (symmetric and asymmetric) and read SNM free 6T (or 8T) SRAM bitcells for 32 nm technology node with NBTI effect and the duty cycle $\beta = 0.25$ for 5 years of time span. The butterfly curve of a symmetric 6T SRAM bitcell is almost symmetric and it has negligible effect of shifted threshold voltage due to NBTI. For asymmetric

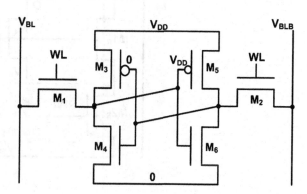

Fig. 6.6 Schematic diagram of a symmetric standard 6T SRAM bitcell

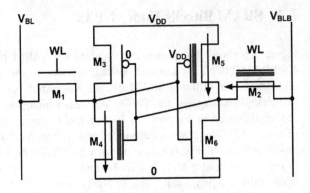

Fig. 6.7 Asymmetric 6T SRAM bitcell for reduced leakage current based on dual-V_{TH} transistors (*shaded transistors are high* V_{TH})

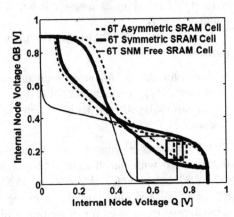

Fig. 6.8 Different SRAM bitcells *butterfly curves* for read SNM measurement under NBTI degradation

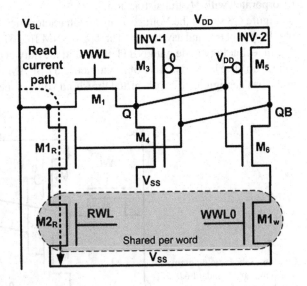

Fig. 6.9 Read SNM free 6T SRAM bitcell with shared read and write assist transistors (*shaded*) per word

Table 6.1 Read SNM and WNM degradation due to NBTI for 32 nm technology at 125°C

Cells	NBTI	Hold SNM [mV]	Read SNM [mV]	WNM [mV]
6T symmetric	t = 0	267.01	135.15	160
SRAM	t = 5 Years	247.15	120.28	179
6T asymmetric	t = 0	245.35	101.02	140
SRAM	t = 5 Years	224.09	86.36	170
6T SNM free	t = 0	267.01	267.01	160
SRAM	t = 5 Years	247.15	247.15	179

6T SRAM bitcell, butterfly curve is not symmetric that is because of dual-V_{TH} devices. However, shifted threshold voltage due to NBTI of transistor M_3 has less effect as compared to high-V_{TH} devices (M_2, M_4 and M_5) used in the bitcell. In other words, low V_{TH} devices age faster than the high V_{TH} devices and NBTI V_{TH} degradation is more significant at elevated temperature. For the read SNM free 6T (or 8T) SRAM bitcell, read SNM is equivalent to hold SNM or in other words an isolated read port held the data storage nodes unchanged. Hence, read SNM free 6T (or 8T) SRAM bitcell has better read SNM and 6T asymmetric SRAM bitcell has worst read SNM under NBTI.

The stability parameters (SNM and WNM) of 6T (symmetric and asymmetric) and read SNM free 6T (or 8T) SRAM bitcells are analyzed using HSPICE simulation in order to investigate the NBTI effects. Under NBTI pull-up (PMOS) transistors are more weakened, which skews the transfer characteristics of the inverter as a result degraded hold and read SNM. The write noise margin (WNM) of a standard 6T SRAM bitcell may improve or degrade depending upon the probability of stress. By weakening of the pull-up (PMOS) transistors due to NBTI may improve the WNM allowing '0' to be written more easily, on the other hand a small noise may flip the bitcell content quickly.

The Static Noise Margin (SNM) obtained from the butterfly curve is used for measuring the read and hold stability. The SNM is estimated graphically as the length of a side of the largest square that can be embedded inside the lobes of the butterfly curve [95]. While the write stability (WNM) is measured using the write trip point, defined as the minimum amount of voltage needed on the bitline to flip the bitcell content [35].

Tables 6.1 and 6.2 show the degradation and recovery in stability parameters, respectively, for the 6T (symmetric and asymmetric) and read SNM free 6T (or 8T) SRAM bitcell caches. Degradation in read SNM and WNM are calculated by simulating the SRAM bitcells for the shifted value of threshold voltage due to NBTI after 5 years of time for different duty cycles. Elevated temperature 125°C has been considered for the simulation as NBTI is more pronounced at higher temperature. The percentage of recovery in SNM and WNM are calculated by incorporating the dynamically recovered threshold voltage for $\beta = 0.25$ in the SRAM bitcells, simulation results for different SRAM bitcells are tabulated in Table 6.2.

Table 6.2 Recovery of read SNM and WNM for 32 nm technology at 125°C and duty cycle $(\beta) = 0.25$

Cells	NBTI	Hold SNM [mV]	Read SNM [mV]	WNM [mV]
6T symmetric	$\beta = 0.25$	260.19	130.14	160
SRAM	Recovery(%)	66.13	66.30	65.51
6T asymmetric	$\beta = 0.25$	238.06	96.00	151
SRAM	Recovery(%)	65.70	65.89	63.33
6T SNM free	$\beta = 0.25$	260.19	260.19	160
SRAM	Recovery(%)	66.13	66.13	65.51

The amount of degradation in the read SNM and WNM is approximately 10% for both symmetric and asymmetric 6T SRAM bitcells after 5 years of time span. While there is a 28.3% reduction in read SNM for the asymmetric 6T SRAM bitcell as compared to symmetric 6T SRAM bitcell, as shown in Table 6.1. This drastic reduction in read SNM is mainly due to asymmetry introduced by the dual-V_{TH} devices used in the bitcell, in order to minimize the leakage current. Also an opposite trend of WNM is observed, WNM of the initial (t = 0) SRAM bitcell is lower than the stressed (t = 5 years) SRAM bitcell, because of increase in trip-point of inverter (M3, and M4). Therefore, higher voltage is needed at the bitlines to write into the SRAM bitcells due to aging effect.

Recovery in the read SNM for different SRAM bitcells are almost equal, as shown in Table 6.2. However, recovery in WNM for asymmetric 6T SRAM bitcell is slightly higher than the symmetric 6T SRAM bitcell. Since, asymmetric SRAM bitcell consists of high V_{TH} devices, which has less impact of NBTI as compared to low V_{TH} devices used in symmetric SRAM bitcell, hence, higher recovery in WNM is observed.

6.5 Leakage Energy Saving Techniques in Caches

With increasing device density, lower supply voltage and threshold voltage, the trend of energy optimization is shifted from dynamic to leakage energy. Leakage energy is dominating in the dense cache memories that occupy a major portion of a die. Many techniques have been proposed in the past to reduce leakage during idle mode by switching off the supply voltage, such as, gated-V_{DD} to dynamically shutdown cache blocks [54, 118] and used in conjunction with software to remove the dead objects [26]. However, dynamically shutting down the cache blocks results in state of the cache memory being lost (state-destroying). Various alternate state-preserving techniques have also been proposed in the recent past for leakage savings [70, 106]. The choice between state-preserving and state-destroying depends on the additional overheads needed for restoring the lost state from the other level of memory hierarchy.

By lowering the supply voltage dynamically (gated-V_{DD}) of the cache blocks, leakage energy can be saved. If the path to V_{DD} or ground is completely cut-off then the state of the cache is lost and this is considered as a state destroying mode. Otherwise, if the supply voltage is lowered but the SRAM bitcells are still able to maintain their state then this is considered as a state preserving mode and the cache is termed to be in a sleep state. An earlier work [69] had considered a number of state preserving and state destroying strategies for leakage savings and focused on exploiting the data duplication present in an on-chip L1-L2 cache hierarchy (which consists of an L1 instruction cache, an L1 data cache, and a unified L2 cache). In general, in an L1-L2 cache hierarchy, the data present in L1 is also contained in L2, hence, leakage energy can be saved by keeping only one active copy of the data. The five main strategies, that exploit the state-preserving and state-destroying leakage optimization mechanisms are as follows [69]:

- CONSERVATIVE: When a block in L1 is written to, then the corresponding sub-block in L2 is fully turned off in a state destroying mode. Since the block in L1 is dirty then the block in L2 is dead and can be safely deactivated. Since, instructions are not written to, this strategy cannot be optimized.
- SPECULATIVE I: When a block is brought from L2 to L1, the block in L1 is put in a state-preserving mode immediately. It does not wait for the L1 block to become dirty nor does it lose data. If the evicted block had become dirty then the block in L2 is reactivated and written into.
- SPECULATIVE II: Similar to Speculative I but instead the L2 block is put into the state destroying mode instead of sleep mode.
- SPECULATIVE III: This is similar to Speculative I, but the block in L2 is speculatively woken up when the L1 block is evicted.
- SPECULATIVE IV: This is similar to Speculative I, except the L2 block is reactivated and written back, whenever the corresponding L1 cache block needs to be replaced.

6.5.1   Leakage Energy Saving Cache Configurations

Media and array dominated application benchmarks such as Media-Bench, Spec and Perfect Club are used for cycle-accurate simulations, under different power saving cache strategies to determine the leakage energy savings. The leakage energy saved under these cache strategies was used to calculate the average inactive time experienced by the cache blocks. Based on the average inactive time, the duty cycles (β) are derived and following different cases (cache configurations) are formed:

- CASE 1: Exploits the skewed distribution of '0' with average occupancy of 75% of the time. Due to the symmetric nature of SRAMs if one of the PMOSs has a duty cycle of 0.75 then the other has a duty cycle of 0.25.

Table 6.3 Drived duty cycles and resulting change (or shift) in threshold voltage for different cases (or cache configurations)

Cases	β	Description	PMOS 32 nm/45 nm $\Delta\lvert V_{TH}\rvert$ [mV]
1	0.750	Average occupancy of 0 in cache	47.7/54.12
2	0.500	Periodic bit flipping	36.2/41.1
3	0.425	Periodic bit flipping + disable	33.6/38.13
4	0.385	Periodic bit flipping + sleep mode + speculative wake up	32.2/36.6
5	0.315	Periodic bit flipping + sleep mode + on demand wake up	29.5/33.86
6	0.250	Periodic bit flipping + sleep mode + decay to state destroying	27.5/31.25

Table 6.4 PTM Technology model parameters used for simulation of different SRAM bitcells and cache configurations

Tech.	V_{DD} [V]	V_{TH} [V]	High V_{TH} [V]	T_{OX} [nm]
32 nm	0.9	0.16	0.24	1
45 nm	1	0.18	0.27	1.1

- CASE 2: Under normal cache with an additional record stating if the value is inverted or not, or with periodic inversions the duty cycle will approach to 0.5 equalizing the degradation on both the PMOS transistors.
- CASE 3: Extends CASE 2 and incorporates the Conservative strategy of disabling cache blocks.
- CASE 4: Employs sleep mode of cache blocks when the block is written into L1. It attempts to speculatively wake up the L2 block before it is needed.
- CASE 5: It is similar to CASE 4, but it only wakes up the block when it is needed.
- CASE 6: It combines the sleep mode with the disabled mode using a timer that if it expires switches the cache block-off instead of sleeping.

Table 6.3 summarizes different cases formed on the basis of state-preserving and state-destroying leakage saving strategies discussed above, and their corresponding duty cycles in column 2. The NBTI induced shift in threshold voltage ($\Delta\lvert V_{TH}\rvert$) for 32 nm and 45 nm technology node PMOS transistors after 5 years is estimated using the model described in Eqs. 6.1, considering their respective duty cycles (β). We use PTM 32 nm and 45 nm technology models [88], with following parameters provided in the Table 6.4.

The PMOS transistor models with shifted threshold voltage due to NBTI for different duty cycles are incorporated in HSPICE net list to simulate different SRAM bitcell based cache configurations for their stability parameters, leakage current measurement and process variation analysis.

6.6 Stability Recovery Under Different Cache Configurations

In order to demonstrate the effectiveness of dynamically recovering the change in threshold voltage (V_{TH}) under different cache configurations, formed on the basis of leakage energy saving cache strategies, recovery in stability parameters read SNM and WNM is measured for different SRAM bitcells.

6.6.1 Read SNM Recovery

As it is shown in the previous Chaps. 1 and 2 that the read SNM is one of the critical stability parameters of a SRAM bitcell in nano-regime. It is crucial for a successful implementation of the reliable and high performance caches. The impact of NBTI due to aging makes it more challenging. Figures 6.10 and 6.11 show the recovery in read SNM for 32 nm and 45 nm technology nodes respectively, at 70°C and 125°C for different SRAM bitcells based cache configurations. Recovery in read SNM for different SRAM cache configurations varies from 38% for cache

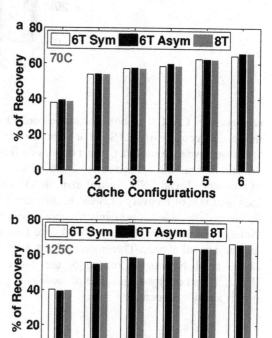

Fig. 6.10 Percentage (%) of recovery of read static noise margin (SNM) for different cache configurations based on 32 nm node SRAM bitcells (a) Recovery in read SNM at 70°C (b) Recovery in read SNM at 125°C

Fig. 6.11 Percentage (%) of recovery of read static noise margin (SNM) for different cache configurations based on 45 nm node SRAM bitcells. (a) Recovery in read SNM at 70°C. (b) Recovery in read SNM at 125°C

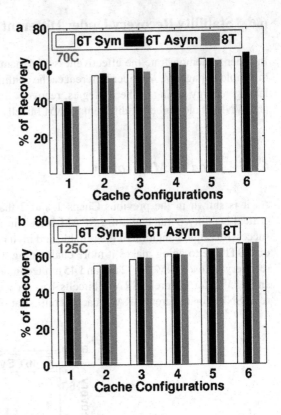

configuration CASE-1 to 66% for cache configuration CASE-6. There is a slight increase in recovery of read SNM at higher temperature.

As we have seen in Sect. 6.2 that duty cycle β has significant role in modulating the shift in threshold voltage due to aging, which is also visible from the percentage of recovery of read SNM for different SRAM bitcell based cache configurations. Increased rate of recovery of SNM in different SRAM caches is specifically seen for lower duty cycles. It is mainly because of less impact of NBTI under lower duty cycles, or in other words, cache blocks are kept in the state-destroying mode for regular interval of time. Therefore, CASE-6 has least impact of NBTI or has better capability of dynamically recovering the shifted threshold voltage due to NBTI. Hence, it can be a good candidate of cache configuration, where reliability and life span is a major concern.

Fig. 6.12 Percentage (%) of recovery of write noise margin (WNM) for different cache configurations based on 32 nm node SRAM bitcells. (**a**) Recovery in write noise margin (WNM) at 70°C. (**b**) Recovery in write noise margin (WNM) at 125°C

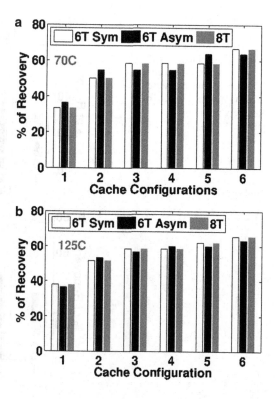

6.6.2   WNM Recovery

A similar trend has also been observed in the recovery of WNM for different SRAM bitcells based cache configurations. However, rate of recovery of WNM for asymmetric 6T SRAM bitcell caches is slightly lower than the symmetric 6T SRAM bitcells based caches, as shown in Figs. 6.12 and 6.13 for 32 nm and 45 nm nodes, respectively. While the rate of recovery of WNM of the SNM free 6T (or 8T) SRAM has almost equivalent to symmetric 6T SRAM, since write operation in the SNM free 6T (or 8T) SRAM will takes place in similar fashion of symmetric 6T, assuming that the regular V_{TH} devices are used.

Increased rate of recovery of WNM is also observed, in particular for lower duty cycles in different SRAM bitcells based cache configurations. It is purely because of lower duty cycles. It can be seen from Figs. 6.12 and 6.13, that the CASE-6 has least impact of NBTI or in other words it has better capability of dynamically recovering the shifted threshold voltage due to NBTI.

Fig. 6.13 Percentage (%) of recovery of write noise margin (WNM) for different cache configurations based on 45 nm node SRAM bitcells. (**a**) Recovery in write noise margin (WNM) at 70°C. (**b**) Recovery in write noise margin (WNM) at 125°C

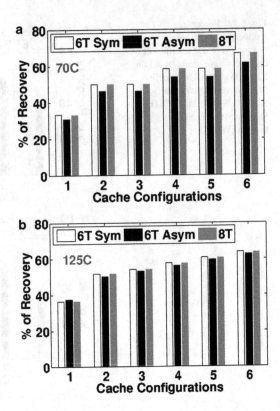

6.7   Effect of NBTI Under Process Variation

Increasing sensitivity of variation in design and process parameters, particularly threshold voltage leads to a greater loss of parametric yield with respect to SRAM bitcell noise margins or stability parameters [10]. The effect of NBTI has direct impact on the PMOS device threshold voltage as a result SRAM bitcells may be more susceptible to parametric failure due to aging effect. As it is evident from the previous section that read SNM and WNM follow almost the similar trend, hence, the study of process variations on leakage current is considered in this section for different cache configurations. Leakage current has strong exponential dependence on threshold voltage. In order to investigate the effect of NBTI along with process variations on read SNM and leakage current, 1,000 Monte Carlo simulations are preformed for each cache configuration, it is assumed that a 15% variation in V_{TH} with 3σ as an independent random variable for all the transistors in 6T and SNM free 6T (or 8T) SRAM bitcells with Gaussian distribution.

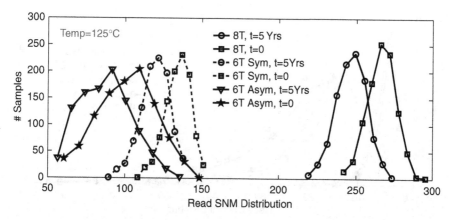

Fig. 6.14 Read SNM distribution of different SRAM bitcells for 32 nm technology node

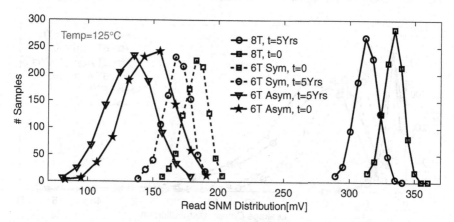

Fig. 6.15 Read SNM distribution of different SRAM bitcells for 45 nm technology node

6.7.1   Read SNM Distribution Under Process Variation

Figures 6.14 and 6.15 show the read SNM distribution of 32 nm and 45 nm technology nodes, respectively, at 125°C temperature for different SRAM bitcells. Degradation in mean read SNM due to NBTI after 5 years of time is clearly visible. The effect of process variations along with NBTI is more dominant in 32 nm node as compared to 45 nm technology node, which is quite obvious and expected, since smaller feature sized devices are more susceptible to process variations. However, asymmetric 6T SRAM bitcell shows large standard deviation in read SNM as compared to its counterpart symmetric 6T and the SNM free 6T (or 8T) SRAM bitcells for both the technology nodes. Furthermore, higher mean value of read SNM in the SNM free 6T (or 8T) SRAM bitcell is achieved, as shown in Figs. 6.14 and 6.15, this improvement in read SNM is mainly due to the isolated read port.

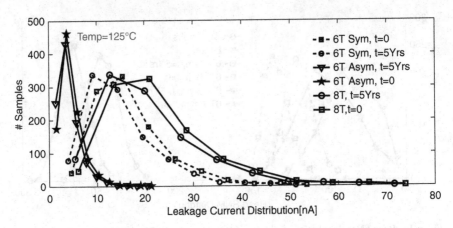

Fig. 6.16 Leakage current distribution of different SRAM bitcells for 32 nm technology node

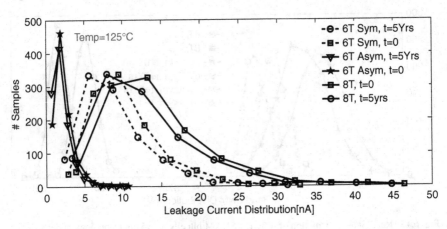

Fig. 6.17 Leakage current distribution of different SRAM bitcells for 45 nm technology node

6.7.2   Leakage Current Distribution Under Process Variations

Figures 6.16 and 6.17 show the leakage current distribution of 32 nm and 45 nm
technology nodes respectively at 125°C temperature for different SRAM bitcell
configurations. It is seen from Figs. 6.16 and 6.17 that the leakage current follows
the log-normal distribution because of its exponential dependence with the threshold
voltage. Also there is significant reduction in mean leakage current for asymmetric
6T SRAM bitcell as compared to symmetric 6T and the SNM free 6T (or 8T)
bitcells, this is due to dual-V_{TH} devices used in the asymmetric 6T SRAM bitcell.
However, the SNM free 6T (or 8T) bitcell has higher mean leakage current than 6T
(symmetric and asymmetric), since it has an extra leakage path from the read port.

6.8 Summary

In this chapter, a detailed examination of the effect of NBTI along with process variations on different SRAM bitcells in low power cache configurations, is presented. The impact of NBTI is studied for different SRAM bitcells namely the 6T (symmetric and asymmetric) and the read SNM free 6T with 32 nm and 45 nm technology nodes at different temperatures. It is observed that the combination of sleeping modes with complete shut-off of the cache when not required (i.e. $\beta = 0.25$) provides the best recovery of $\sim$66%. Hence, this cache configuration is a good candidate for applications where reliability and life span is a major concern. Also the use of 6T asymmetric bitcells is not advisable only just because of saving in leakage energy but due to poor read SNM. In particular, for reliable applications because of lowest read SNM and higher susceptibility to process variation. However, the read SNM free 6T SRAM bitcell based cache configurations are best suited for reliable applications, since, this bitcell is less vulnerable to process variations and yields the higher read SNM.

References

1. Agarwal, K., Nassif, S.: Statistical analysis of SRAM cell stability. In: DAC '06: Proceedings of the 43rd Annual Conference on Design Automation, San Francisco, pp. 57–62. ACM Press, New York (2006)
2. Agarwal, A., Li, H., Roy, K.: A single-vt low-leakage gated-ground cache for deep submicron. IEEE J. Solid-State Circuit **38**(2), 319–328 (2003)
3. Alam, M.A., Mahapatra, S.: A comprehensive model of PMOS NBTI degradation. Microelectron. Reliab. **45**(1), 71-81 (2005). doi:10.1016/j.microrel. 2004.03.019.http://www.sciencedirect.com/science/article/B6V47-4D0Y2H7-1/2/ 7717df9170a06e98ea4d727d74b6c 2e0
4. Aly, R., Bayoumi, M.: Low-power cache design using 7T SRAM cell. IEEE Trans. Circuit Syst. II. Express Briefs **54**(4), 318–322 (2007)
5. Amelifard, B., Fallah, F., Pedram, M.: Leakage minimization of SRAM cells in a dual- and dual- technology. IEEE Trans. Very Large Scale Integr. Syst. **16**(7), 851–860 (2008). doi:10. 1109/TVLSI.2008.2000459
6. Anis, M., Areibi, S., Elmasry, M.: Design and optimization of multithreshold CMOS (MTCMOS) circuits. IEEE Trans. Comput. Aided Des. Integr. Circuit Syst. **22**(10), 1324-1342 (2003)
7. Arnaud, F., Boeuf, F., Salvetti, F., Lenoble, D., Wacquant, F., Regnier, C., Morin, P., Emonet, N., Denis, E., Oberlin, J., Ceccarelli, D., Vannier, P., Imbert, G., Sicard, A., Perrot, C., Belmont, O., Guilmeau, I., Sassoulas, P., Delmedico, S., Palla, R., Leverd, F., Beverina, A., DeJonghe, V., Broekaart, M., Pain, L., Todeschini, J., Charpin, M., Laplanche, Y., Neira, D., Vachellerie, V., Borot, B., Devoivre, T., Bicais, N., Hinschberger, B., Pantel, R., Revil, N., Parthasarathy, C., Planes, N., Brut, H., Farkas, J., Uginet, J., Stolk, P., Woo, M.: A functional 0.69 mu;m2 embedded 6T-SRAM bit cell for 65 nm CMOS platform. In: Symposium on VLSI Technology, 2003. Digest of Technical Papers, pp. 65–66 (2003). doi:10.1109/VLSIT.2003.1221088
8. Azizi, N., Najm, F., Moshovos, A.: Low-leakage asymmetric-cell SRAM. IEEE Trans. Very Large Scale Integr. Syst. **11**(4), 701–715 (2003). doi:10.1109/TVLSI.2003.816139
9. Ball, M., Rosal, J., McKee, R., Loh, W., Houston, T., Garcia, R., Raval, J., Li, D., Hollingsworth, R., Gury, R., Eklund, R., Vaccani, J., Castellano, B., Piacibello, F., Ashburn, S., Tsao, A., Krishnan, A., Ondrusek, J., Anderson, T.: A screening methodology for VMIN drift in SRAM arrays with application to sub-65 nm nodes. In: International Electron Devices Meeting, 2006. IEDM '06, pp. 1–4 (2006). doi:10.1109/IEDM.2006.346883
10. Bhavnagarwala, A.J., Tang, X., Meindl, J.D.: The impact of intrinsic device fluctuations on CMOS SRAM cell stability. IEEE J. Solid-State Circuit **36**, 658-665 (2001)

11. Bhavnagarwala, A., Kosonocky, S., Kowalczyk, S., Joshi, R., Chan, Y., Srinivasan, U., Wadhwa, J.: A transregional CMOS SRAM with single, logic vdd and dynamic power rails. In: Symposium on VLSI Circuits, 2004. Digest of Technical Papers, Honolulu, pp. 292–293 (2004)

12. Bhavnagarwala, A., Kosonocky, S., Radens, C., Stawiasz, K., Mann, R., Ye, Q., Chin, K.: Fluctuation limits amp; scaling opportunities for CMOS SRAM cells. In: IEEE International Electron Devices Meeting, 2005. IEDM Technical Digest, Washington, pp. 659–662 (2005)

13. Bhavnagarwala, A., Kosonocky, S., Radens, C., Chan, Y., Stawiasz, K., Srinivasan, U., Kowalczyk, S., Ziegler, M.: A sub-600-mv, fluctuation tolerant 65-nm CMOS SRAM array with dynamic cell biasing. IEEE J. Solid-State Circuit **43**(4), 946–955 (2008). doi:10.1109/JSSC.2008.917506

14. Bhuwalka, K., Sedlmaier, S., Ludsteck, A., Tolksdorf, C., Schulze, J., Eisele, I.: Vertical tunnel field-effect transistor. IEEE Trans. Electron Device **51**(2), 279–282 (2004). doi:10.1109/TED.2003.821575

15. Boeuf, F., Arnaud, F., Boccaccio, C., Salvetti, F., Todeschini, J., Pain, L., Jurdit, M., Manakli, S., Icard, B., Planes, N., Gierczynski, N., Denorme, S., Borot, B., Ortolland, C., Duriez, B., Tavel, B., Gouraud, P., Broekaart, M., Dejonghe, V., Brun, P., Guyader, F., Morini, P., Reddy, C., Aminpur, M., Laviron, C., Smith, S., Jacquemin, J., Mellier, M., Andre, F., Bicais-Lepinay, N., Jullian, S., Bustos, J., Skotnicki, T.: 0.248 mu;m2 and 0.334 mu;m2 conventional bulk 6T-SRAM bit-cells for 45 nm node low cost – general purpose applications. In: Symposium on VLSI Technology, 2005. Digest of Technical Papers, pp. 130–131 (2005). doi:10.1109/.2005.1469240

16. Borkar, S.: Design challenges of technology scaling. IEEE Micro **19**(4), 23–29 (1999). doi:10.1109/40.782564

17. Calhoun, B., Chandrakasan, A.: Static noise margin variation for sub-threshold SRAM in 65-nm CMOS. IEEE J. Solid-State Circuit **41**(7), 1673–1679 (2006). doi:10.1109/JSSC.2006.873215

18. Calhoun, B.H., Chandrakasan, A.P.: A 256-kb 65-nm sub-threshold SRAM design for ultralow-voltage operation. IEEE J. Solid-State Circuit **42**(3), 680–688 (2007)

19. Calhoun, B., Daly, D., Verma, N., Finchelstein, D., Wentzloff, D., Wang, A., Cho, S.H., Chandrakasan, A.: Design considerations for ultra-low energy wireless microsensor nodes. IEEE Trans. Comput. **54**(6), 727–740 (2005). doi:10.1109/TC.2005.98

20. Carlson, I., Andersson, S., Natarajan, S., Alvandpour, A.: A high density, low leakage, 5T SRAM for embedded caches. In: Proceeding of the 30th European Solid-State Circuits Conference, ESSCIRC 2004, Leuven, pp. 215–218 (2004)

21. Chang, Y.J., Lai, F.: Dynamic zero-sensitivity scheme for low-power cache memories. IEEE Micro **25**(4), 20–32 (2005). doi:10.1109/MM.2005.64

22. Chang, H., Sapatnekar, S.: Full-chip analysis of leakage power under process variations, including spatial correlations. In: Proceedings of the 42nd Design Automation Conference 2005, pp. 523–528 (2005). doi:10.1109/DAC.2005.193865

23. Chang, L., Fried, D., Hergenrother, J., Sleight, J., Dennard, R., Montoye, R., Sekaric, L., McNab, S., Topol, A., Adams, C., Guarini, K., Haensch, W.: Stable SRAM cell design for the 32 nm node and beyond. In: Symposium on VLSI Technology, 2005. Digest of Technical Papers, Kyoto, pp. 128–129. 14–16 June 2005

24. Chang, L., Nakamura, Y., Montoye, R., Sawada, J., Martin, A., Kinoshita, K., Gebara, F., Agarwal, K., Acharyya, D., Haensch, W., Hosokawa, K., Jamsek, D.: A 5.3 ghz 8T-SRAM with operation down to 0.41 v in 65 nm CMOS. In: IEEE Symposium on VLSI Circuits, 2007, Kyoto, pp. 252–253 (2007)

25. Chang, L., Montoye, R., Nakamura, Y., Batson, K., Eickemeyer, R., Dennard, R., Haensch, W., Jamsek, D.: An 8T-SRAM for variability tolerance and low-voltage operation in high-performance caches. IEEE J. Solid-State Circuit **43**(4), 956–963 (2008)

26. Chen, G., Shetty, R., Kandemir, M., Vijaykrishnan, N., Irwin, M., Wolczko, M.: Tuning garbage collection in an embedded java environment. In: Proceedings of the Eighth International Symposium on High-Performance Computer Architecture, 2002, Cambridge, pp. 92–103 (2002)

27. Chen, G., Chuah, K., Li, M., Chan, D., Ang, C., Zheng, J., Jin, Y., Kwong, D.: Dynamic NBTI of PMOS transistors and its impact on device lifetime. In: IEEE International Reliability Physics Symposium Proceedings, 2003. 41st Annual, Dallas, pp. 196–202 (2003)
28. Chen, G.K., Blaauw, D., Mudge, T., Sylvester, D., Kim, N.S.: Yield-driven near-threshold SRAM design. In: ICCAD '07: Proceedings of the 2007 IEEE/ACM International Conference on Computer-Aided Design, Lyon, pp. 660–666. IEEE Press, Piscataway (2007)
29. Cragon, H.G.: Memory Systems and Pipelined Processors, Chapter 1. Jones and Barlett, Sudbury (1996)
30. Dennard, R.H.: Field-effect transistor memory. US Patent No. 3387286 (1968)
31. Enomoto, T., Oka, Y., Shikano, H.: A self-controllable voltage level (svl) circuit and its low-power high-speed cmos circuit applications. IEEE J. Solid-State Circuit **38**(7), 1220–1226 (2003)
32. Fair, R., Wivell, H.: Zener and avalanche breakdown in As-implanted low-voltage Si n-p junctions. IEEE Trans. Electron Devices **23**(5), 512–518 (1976)
33. Fischer, T., Amirante, E., Huber, P., Nirschl, T., Olbrich, A., Ostermayr, M., Schmitt-Landsiedel, D.: Analysis of read current and write trip voltage variability from a 1-MB SRAM test structure. IEEE Trans. Semicond. Manuf. **21**(4), 534–541 (2008). doi:10.1109/TSM.2008.2004329
34. Gierczynski, N., Borot, B., Planes, N., Brut, H.: A new combined methodology for write-margin extraction of advanced SRAM. In: IEEE International Conference on Microelectronic Test Structures 2007, ICMTS '07, pp. 97–100 (2007). doi:10.1109/ICMTS.2007.374463
35. Grossar, E., Stucchi, M., Maex, K., Dehaene, W.: Read stability and write-ability analysis of SRAM cells for nanometer technologies. IEEE J. Solid-State Circuit **41**(11), 2577–2588 (2006). doi:10.1109/JSSC.2006.883344
36. Jiajing W., Nalam, S., Calhoun, B.H.: Analyzing static and dynamic write margin for nanometer SRAMs, ACM/IEEE International Symposium on Low Power Electronics and Design (ISLPED), 129–134 (2008) doi: 10.1145/1393921.1393954. http://ieeexplore.ieee.org/stamp/stamp.jsp?tp=&arnumber=5529055&isnumber=5529013
37. Guo, Z., Carlson, A., Pang, L.T., Duong, K., Liu, T.J.K., Nikolic, B.: Large-scale read/write margin measurement in 45 nm CMOS SRAM arrays. In: IEEE Symposium on VLSI Circuits, 2008, pp. 42–43 (2008). doi:10.1109/VLSIC.2008.4585944
38. Guo, Z., Carlson, A., Pang, L.T., Duong, K., Liu, T.J.K., Nikolic, B.: Large-scale SRAM variability characterization in 45 nm CMOS. IEEE J. Solid-State Circuit **44**(11), 3174–3192 (2009). doi:10.1109/JSSC.2009.2032698
39. Gupta, V., Anis, M.: Statistical design of the 6T SRAM bit cell. IEEE Trans. Circuit Syst. I. Regul. Pap. **57**(1), 93–104 (2010). doi:10.1109/TCSI.2009.2016633
40. Heald, R., Wang, P.: Variability in sub-100 nm SRAM designs. In: International Conference on Computer Aided Design, 2004. ICCAD-2004, San Jose, pp. 347–352 (2004)
41. Heald, R., Wang, P.: Variability in sub-100 nm SRAM designs. In: IEEE/ACM International Conference on Computer Aided Design, San Jose, pp. 347–352 (2004)
42. Hennessy, J.L., Patterson, D.: Computer Architecture: A Quantitative Approach, Chapter 5. Morgan Kaufman, San Francisco (2006)
43. Hirose, T., Kuriyama, H., Murakami, S., Yuzuriha, K., Mukai, T., Tsutsumi, K., Nishimura, Y., Kohno, Y., Anami, K.: A 20 ns 4 mb CMOS SRAM with hierarchical word decoding architecture. In: IEEE International Solid-State Circuits Conference, Digest of Technical Papers. 37th ISSCC 1990, pp. 132–133 (1990). doi:10.1109/ISSCC.1990.110162
44. Hobson, R.: A new single-ended SRAM cell with write-assist. IEEE Trans. Very Large Scale Integr. Syst. **15**(2), 173–181 (2007)
45. Hurkx, G., Klaassen, D., Knuvers, M.: A new recombination model for device simulation including tunneling. IEEE Trans. Electron Devices **39**(2), 331–338 (1992). doi:10.1109/16.121690
46. Ieong, M., Solomon, P., Laux, S., Wong, H.S., Chidambarrao, D.: Comparison of raised and schottky source/drain mosfets using a novel tunneling contact model. In: International Electron Devices Meeting, 1998. IEDM '98 Technical Digest, pp. 733–736 (1998). doi:10.1109/IEDM.1998.746461

47. Itoh, K., Sasaki, K., Nakagome, Y.: Trends in low-power ram circuit technologies. Proc. IEEE **83**(4), 524–543 (1995). doi:10.1109/5.371965
48. ITRS: International technology road map for semiconductors, test and test equipments. http://public.itrs.net/ (2006)
49. Kaffashian, M.H., Lotfi, R., Mafinezhad, K., Mahmoodi, H.: Impact of NBTI on performance of domino logic circuits in nano-scale CMOS. Microelectron. J. **42**(12), 1327–1334 (2011). doi:10.1016/j.mejo.2011.09.009. http://www.sciencedirect.com/science/article/pii/S0026269211001984
50. Kang, K., Kufluoglu, H., Alain, M., Roy, K.: Efficient transistor-level sizing technique under temporal performance degradation due to NBTI. In: International Conference on Computer Design, 2006. ICCD 2006, pp. 216–221 (2006). doi:10.1109/ICCD.2006.4380820
51. Kao, J., Chandrakasan, A., Antoniadis, D.: Transistor sizing issues and tool for multi-threshold CMOS technology. In: Proceedings of the 34th Design Automation Conference, 1997, Anaheim, pp. 409–414 (1997)
52. Kao, J., Narendra, S., Chandrakasan, A.: MTCMOS hierarchical sizing based on mutual exclusive discharge patterns. In: DAC '98: Proceedings of the 35th Annual Conference on Design Automation, San Francico, pp. 495–500 (1998)
53. Kawaguchi, H., Kanda, K., Nose, K., Hattori, S., Dwi, D., Antono, D., Yamada, D., Miyazaki, T., Inagaki, K., Hiramoto, T., Sakurai, T.: A 0.5 v, 400 mhz, v00-hopping processor with zero-vth fd-soi technology. In: IEEE International Solid-State Circuits Conference, Digest of Technical Papers. ISSCC, 2003, vol. 1, pp. 106–481 (2003). doi:10.1109/ISSCC.2003.1234227
54. Kaxiras, S., Hu, Z., Martonosi, M.: Cache decay: exploiting generational behavior to reduce cache leakage power. In: Proceedings of the 28th Annual International Symposium on Computer Architecture, 2001, pp. 240–251 (2001). doi:10.1109/ISCA.2001.937453
55. Khalil, D., Khellah, M., Kim, N.S., Ismail, Y., Karnik, T., De, V.: Accurate estimation of SRAM dynamic stability. IEEE Trans. Very Large Scale Integr. Syst. **16**(12), 1639–1647 (2008). doi:10.1109/TVLSI.2008.2001941
56. Khare, M., Ku, S., Donaton, R., Greco, S., Brodsky, C., Chen, X., Chou, A., DellaGuardia, R., Deshpande, S., Doris, B., Fung, S., Gabor, A., Gribelyuk, M., Holmes, S., Jamin, F., Lai, W., Lee, W., Li, Y., McFarland, P., Mo, R., Mittl, S., Narasimha, S., Nielsen, D., Purtell, R., Rausch, W., Sankaran, S., Snare, J., Tsou, L., Vayshenker, A., Wagner, T., Wehella-Gamage, D., Wu, E., Wu, S., Yan, W., Barth, E., Ferguson, R., Gilbert, P., Schepis, D., Sekiguchi, A., Goldblatt, R., Welser, J., Muller, K., Agnello, P.: A high performance 90 nm SOI technology with 0.992 m2 6T-SRAM cell. In: International Electron Devices Meeting, 2002. IEDM '02. Digest, pp. 407–410 (2002). doi:10.1109/IEDM.2002.1175865
57. Kim, N.S., Flautner, K., Blaauw, D., Mudge, T.: Circuit and microarchitectural techniques for reducing cache leakage power. IEEE Trans. Very Large Scale Integr. Syst. **12**(2), 167–184 (2004)
58. Kim, T.H., Liu, J., Keane, J., Kim, C.: A 0.2 v, 480 kb subthreshold SRAM with 1 k cells per bitline for ultra-low-voltage computing. IEEE J. Solid-State Circuit **43**(2), 518–529 (2008)
59. Kim, D., Lee, Y., Cai, J., Lauer, I., Chang, L., Koester, S.J., Sylvester, D., Blaauw, D.: Low power circuit design based on heterojunction tunneling transistors (HETTs). In: Proceedings of the 14th ACM/IEEE International Symposium on Low Power Electronics and Design, ISLPED '09, pp. 219–224. ACM, New York (2009). doi:http://doi.acm.org/10.1145/1594233.1594287. http://doi.acm.org/10.1145/1594233.1594287
60. Kimizuka, N., Yamamoto, T., Mogami, T., Yamaguchi, K., Imai, K., Horiuchi, T.: The impact of bias temperature instability for direct-tunneling ultra-thin gate oxide on mosfet scaling. In: Symposium on VLSI Technology, 1999. Digest of Technical Papers, pp. 73–74 (1999). doi:10.1109/VLSIT.1999.799346
61. Kiyoo, I., Katsuro, S., Yoshinobu, N.: Trends in low-power ram circuit technologies. Proc. IEEE **83**, 524–543 (1995)

62. Krishnan, A., Reddy, V., Chakravarthi, S., Rodriguez, J., John, S., Krishnan, S.: NBTI impact on transistor and circuit: models, mechanisms and scaling effects [MOSFETS]. In: IEEE International Electron Devices Meeting, 2003, IEDM '03 Technical Digest, pp. 14.5.1–14.5.4 (2003). doi:10.1109/IEDM.2003.1269296

63. Kulkarni, J., Kim, K., Roy, K.: A 160 mv robust schmitt trigger based subthreshold SRAM. IEEE J. Solid-State Circuit 42(10), 2303–2313 (2007)

64. Kumar, S., Kim, K., Sapatnekar, S.: Impact of nbti on SRAM read stability and design for reliability. In: 7th International Symposium on Quality Electronic Design, 2006. ISQED '06, pp. 6–218 (2006). doi:10.1109/ISQED.2006.73

65. La Rosa, G., Ng, W.L., Rauch, S., Wong, R., Sudijono, J.: Impact of nbti induced statistical variation to SRAM cell stability. In: IEEE International Reliability Physics Symposium Proceedings 2006, 44th Annual, pp. 274–282 (2006). doi:10.1109/RELPHY.2006.251228

66. Lee, S., Sakurai, T.: Run-time voltage hopping for low-power real-time systems. In: Proceedings of the 37th Design Automation Conference 2000, Los Angeles, pp. 806–809 (2000)

67. Lee, D., Kwong, W., Blaauw, D., Sylvester, D.: Analysis and minimization techniques for total leakage considering gate oxide leakage. In: Proceedings of the Design Automation Conference 2003, Anaheim, pp. 175–180 (2003)

68. Leobandung, E., Nayakama, H., Mocuta, D., Miyamoto, K., Angyal, M., Meer, H., Mc-Stay, K., Ahsan, I., Allen, S., Azuma, A., Belyansky, M., Bentum, R.V., Cheng, J., Chidambarrao, D., Dirahoui, B., Fukasawa, M., Gerhardt, M., Gribelyuk, M., Halle, S., Harifuchi, H., Harmon, D., Heaps-Nelson, J., Hichri, H., Ida, K., Inohara, M., Inouc, I., Jenkins, K., Kawamura, T., Kim, B., Ku, S.K., Kumar, M., Lane, S., Liebmann, L., Logan, R., Melville, I., Miyashita, K., Mocuta, A., O'Neil, P., Ng, M.F., Nogami, T., Nomura, A., Norris, C., Nowak, E., Ono, M., Panda, S., Penny, C., Radens, C., Ramachandran, R., Ray, A., Rhee, S.H., Ryan, J., Shinohara, T., Sudo, G., Sugaya, F., Strane, J., Tan, Y., Tsou, L., Wang, L., Wirbeleit, F., Wu, S., Yamashita, T., Yan, H., Ye, Q., Yoneyama, D., Zamdmer, D., Zhong, H., Zhu, H., Zhu, W., Agnello, P., Bukofsky, S., Bronner, G., Crabbe, E., Freeman, G., Huang, S.F., Ivers, T., Kuroda, H., McHerron, D., Pellerin, J., Toyoshima, Y., Subbanna, S., Kepler, N., Su, L.: High performance 65 nm soi technology with dual stress liner and low capacitance SRAM cell. In: Symposium on VLSI Technology, 2005. Digest of Technical Papers, pp. 126–127 (2005). doi:10.1109/.2005.1469238

69. Li, L., Kadayif, I., Tsai, Y.F., Vijaykrishnan, N., Kandemir, M., Irwin, M., Sivasubramaniam, A.: Leakage energy management in cache hierarchies. In: Proceedings of the International Conference on Parallel Architectures and Compilation Techniques, 2002, pp. 131–140 (2002). doi:10.1109/PACT.2002.1106012

70. Li, L., Kadayif, I., Tsai, Y.F., Narayanan, V., Kandemir, M., Irwin, M.J., Sivasubramaniam, A.: Managing leakage energy in cache hierarchies. J. Instruction-Level Parallel. 5, 1–24 (2003)

71. Li, X., Qin, J., Huang, B., Zhang, X., Bernstein, J.: SRAM circuit-failure modeling and reliability simulation with spice. IEEE Trans. Device Mater. Reliab. 6(2), 235–246 (2006). doi:10.1109/TDMR.2006.876568

72. Lin, J., Toh, E., Shen, C., Sylvester, D., Heng, C., Samudra, G., Yeo, Y.: Compact HSPICE model for IMOS device. Electron. Lett. 44(2), 91–92 (2008). doi:10.1049/el:20083116

73. Liu, Z., Kursun, V.: Characterization of a novel nine-transistor SRAM cell. IEEE Trans. Very Large Scale Integr. Syst. 16(4), 488–492 (2008)

74. Mahapatra, S., Kumar, P., Alam, M.: Investigation and modeling of interface and bulk trap generation during negative bias temperature instability of p-MOSFETS. IEEE Trans. Electron Devices 51(9), 1371–1379 (2004). doi:10.1109/TED.2004.833592

75. Mahmoodi, H., Mukhopadhyay, S., Roy, K.: Estimation of delay variations due to random-dopant fluctuations in nanoscale CMOS circuits. IEEE J. Solid-State Circuit 40(9), 1787–1796 (2005)

76. Mann, R.W., Abadeer, W.W., Breitwisch, M.J., Bula, O., Brown, J.S., Colwill, B.C., Cottrell, P.E., Crocco, W.T., Furkay, S.S., Hauser, M.J., Hook, T.B., Hoyniak, D., Johnson, J.M., Lam, C.M., Mih, R.D., Rivard, J., Moriwaki, A., Phipps, E., Putnam, C.S., Rainey, B.A., Toomey, J.J., Younus, M.I.: Ultralow-power SRAM technology. IBM J. Res. Dev. **47**(5.6), 553–566 (2003). doi:10.1147/rd.475.0553

77. Meterelliyoz, M., Kulkarni, J.P., Roy, K.: Thermal analysis of 8-T SRAM for nano-scaled technologies. In: ISLPED '08: Proceeding of the 13th International Symposium on Low Power Electronics and Design, pp. 123–128. ACM, New York (2008). doi:http://doi.acm.org/10.1145/1393921.1393953

78. Mookerjea, S., Datta, S.: Comparative study of si, ge and inas based steep subthreshold slope tunnel transistors for 0.25 v supply voltage logic applications. In: Device Research Conference, 2008, pp. 47–48 (2008). doi:10.1109/DRC.2008.4800730

79. Mookerjea, S., Krishnan, R., Datta, S., Narayanan, V.: On Enhanced Miller Capacitance Effect in Interband Tunnel Transistors. IEEE, Electron Device Letters, **30**(10), 1102–1104 (2009). doi:10.1109/LED.2009.2028907. http://ieeexplore.ieee.org/stamp/stamp.jsp?tp=&arnumber=5232873&isnumber=5263237

80. Moore, G.: Cramming more components onto integrated circuits. Electronics **38**(8), 534–539 (1965)

81. Moshovos, A., Falsafi, B., Najm, F., Azizi, N.: A case for asymmetric-cell cache memories. IEEE Trans. Very Large Scale Integr. Syst. **13**(7), 877–881 (2005). doi:10.1109/TVLSI.2005.850127

82. Mukhopadhyay, S., Mahmoodi, H., Roy, K.: Modeling of failure probability and statistical design of SRAM array for yield enhancement in nanoscaled CMOS. IEEE Trans. Comput. Aided Des. Integr. Circuit Syst. **24**(12), 1859–1880 (2005). doi:10.1109/TCAD.2005.852295

83. Nii, K., Masuda, Y., Yabuuchi, M., Tsukamoto, Y., Ohbayashi, S., Imaoka, S., Igarashi, M., Tomita, K., Tsuboi, N., Makino, H., Ishibashi, K., Shinohara, H.: A 65 nm ultra-high-density dual-port SRAM with 0.71 μm 8T-cell for SoC. In: Symposium on VLSI Circuits, 2006. Digest of Technical Papers, Honolulu, pp. 130–131 (2006)

84. Ohbayashi, S., Yabuuchi, M., Nii, K., Tsukamoto, Y., Imaoka, S., Oda, Y., Yoshihara, T., Igarashi, M., Takeuchi, M., Kawashima, H., Yamaguchi, Y., Tsukamoto, K., Inuishi, M., Makino, H., Ishibashi, K., Shinohara, H.: A 65-nm soc embedded 6T-SRAM designed for manufacturability with read and write operation stabilizing circuits. IEEE J. Solid-State Circuit **42**(4), 820–829 (2007)

85. Ohbayashi, S., Yabuuchi, M., Kono, K., Oda, Y., Imaoka, S., Usui, K., Yonezu, T., Iwamoto, T., Nii, K., Tsukamoto, Y., Arakawa, M., Uchida, T., Okada, M., Ishii, A., Yoshihara, T., Makino, H., Ishibashi, K., Shinohara, H.: A 65 nm embedded SRAM with wafer level burn-in mode, leak-bit redundancy and cu e-trim fuse for known good die. IEEE J. Solid-State Circuit **43**(1), 96–108 (2008)

86. Patterson, D.A., Hennessy, J.L.: Computer Architecture: A Quantitative Approach. Morgan Kaufmann Publishers Inc., San Mateo (1990)

87. Paul, B., Kang, K., Kufluoglu, H., Alam, M., Roy, K.: Impact of nbti on the temporal performance degradation of digital circuits. IEEE Electron Device Lett. **26**(8), 560–562 (2005). doi:10.1109/LED.2005.852523

88. PTM: Predictive technology model. In: Nanoscale Integration and Modeling (NIMO) Group. Arizona State University, Arizona. http://www.eas.asu.edu/ptm/ (2008)

89. Reddick, W.M., Amaratunga, G.A.J.: Silicon surface tunnel transistor. Appl. Phys. Lett. **67**(4), 494–496 (1995). doi:10.1063/1.114547. http://link.aip.org/link/?APL/67/494/1

90. Reddy, V., Krishnan, A., Marshall, A., Rodriguez, J., Natarajan, S., Rost, T., Krishnan, S.: Impact of negative bias temperature instability on digital circuit reliability. In: 40th Annual, Reliability Physics Symposium Proceedings, 2002, pp. 248–254 (2002). doi:10.1109/RELPHY.2002.996644

91. Ricketts, A., Singh, J., Ramakrishnan, K., Vijaykrishnan, N., Pradhan, D.: Investigating the impact of nbti on different power saving cache strategies. In: Design, Automation Test in Europe Conference Exhibition (DATE), 2010, pp. 592–597 (2010). http://dl.acm.org/citation.cfm?id=1870926.1871065

92. Schenk, A.: Rigorous theory and simplified model of the band-to-band tunneling in silicon. Solid-State Electron. **36**(1), 19–34 (1993). doi:10.1016/0038-- 1101(93)90065- X. http://www.sciencedirect.com/science/article/B6TY5--46VC2XP- VV/2/4c4c69fef08ec975219a32ff521d55d6

93. Schroder, D.K., Babcock, J.A.: Negative bias temperature instability: road to cross in deep submicron silicon semiconductor manufacturing. J. Appl. Phys. **94**(1), 1–18 (2003). doi:10. 1063/1.1567461. http://link.aip.org/link/?JAP/94/1/1

94. Seevinck, E., List, F., Lohstroh, J.: Static-noise margin analysis of MOS SRAM cells. J. Solid-State Circuit **25**(2), 784–754 (1987)

95. Seevinck, E., List, F., Lohstroh, J.: Static-noise margin analysis of MOS SRAM cells. IEEE J. Solid-State Circuit **22**(5), 748–754 (1987)

96. Sentaurus, S.: TCAD Sentaurus Device Manual, Release: Z-2007.03. Synopsys (2003)

97. Singh, J., Mathew, J., Pradhan, D., Mohanty, S.: Failure analysis for ultra low power NANO-CMOS SRAM under process variations. In: IEEE International SoC Conference, 2008, pp. 251–254 (2008). doi:10.1109/SOCC.2008.4641522

98. Singh, J., Pradhan, D.K., Hollis, S., Mohanty, S.P., Mathew, J.: Single ended 6T SRAM with isolated read-port for low-power embedded systems. In: Design, Automation and Test in Europe Conference and Exhibition, 2009. DATE '09 (2009) http://ieeexplore.ieee.org/xpls/ abs_all.jsp?arnumber=5090796&tag=1

99. Singh, J., Ramakrishnan, K., Mookerjea, S., Datta, S., Vijaykrishnan, N., Pradhan, D.: A novel Si-tunnel FET based SRAM design for ultra low-power 0.3 v vdd applications. In: Proceedings of the 2010 Asia and South Pacific Design Automation Conference, ASPDAC '10, pp. 181–186. IEEE Press, Piscataway (2010). http://portal.acm.org/citation.cfm?id= 1899721.1899761

100. Suzuki, T., Yamagami, Y., Hatanaka, I., Shibayama, A., Akamatsu, H., Yamauchi, H.: A sub-0.5-v operating embedded SRAM featuring a multi-bit-error-immune hidden-ecc scheme. IEEE J. Solid-State Circuit **41**(1), 152–160 (2006). doi:10.1109/JSSC.2005.859029

101. Suzuki, T., Yamauchi, H., Yamagami, Y., Satomi, K., Akamatsu, H.: A stable 2-port SRAM cell design against simultaneously read/write-disturbed accesses. IEEE J. Solid-State Circuit **43**(9), 2109–2119 (2008)

102. Sylvester, D.: Low power circuit design based on heterojunction tunneling transistors. In: Device Research Conference, Steep Slope or Slippery Slope, Rump Session, pp. 47–48 (2009). doi:10.1109/DRC.2008.4800730

103. Takeda, K., Hagihara, Y., Aimoto, Y., Nomura, M., Nakazawa, Y., Ishii, T., Kobatake, H.: A read-static-noise-margin-free SRAM cell for low-vdd and high-speed applications. IEEE J. Solid-State Circuit **41**(1), 113–121 (2006)

104. Takeuchi, K., Fukai, T., Tsunomura, T., Putra, A., Nishida, A., Kamohara, S., Hiramoto, T.: Understanding random threshold voltage fluctuation by comparing multiple fabs and technologies. In: IEEE International Electron Devices Meeting, 2007, IEDM 2007, pp. 467–470 (2007). http://ieeexplore.ieee.org/xpls/abs_all.jsp?arnumber=4418975

105. Toh, S.O., Guo, Z., Liu, T.J.K., Nikolic, B.: Characterization of dynamic SRAM stability in 45 nm CMOS. IEEE J. Solid-State Circuit **46**(11), 2702–2712 (2011). doi:10.1109/JSSC. 2011.2164300

106. van der Meer, P., van Staveren, A., van Roermund, A.: Ultra-low standby-currents for deep sub-micron vlsi CMOS circuits: smart series switch. In: Proceedings of the ISCAS 2000 Geneva, Circuits and Systems the 2000 IEEE International Symposium on, vol. 4, pp. 1–4 (2000). doi:10.1109/ISCAS.2000.858673

107. Vattikonda, R., Wang, W., Cao, Y.: Modeling and minimization of PMOS NBTI effect for robust nanometer design. In: ACM/IEEE 43rd Design Automation Conference, 2006, pp. 1047–1052 (2006).doi:10.1109/DAC.2006.229436

108. Verma, N., Chandrakasan, A.P.: A 256 kb 65 nm 8T subthreshold SRAM employing sense-amplifier redundancy. IEEE J. Solid-State Circuit **43**(1), 141–149 (2008)

109. Villa, L., Zhang, M., Asanovic, K.: Dynamic zero compression for cache energy reduction. In: International Symposium on Microarchitecture, Monterey, pp. 214–220 (2000)

110. Wang, P.F.: Complementary tunneling FETs (CTFET) in CMOS technology. Ph.D. thesis, TU Munchen, Munich (2003). http://www.ece.udel.edu/~qli

111. Wang, A., Chandrakasan, A.: A 180 mv FFT processor using sub-threshold circuit techniques. In: Proceedings of the IEEE ISSCC Dig. Tech. Papers, pp. 229–293 (2004). http://ieeexplore. ieee.org/xpls/abs_all.jsp?arnumber=1332709

112. Wang, A., Chandrakasan, A.: A 180-mv subthreshold FFT processor using a minimum energy design methodology. IEEE J. Solid-State Circuit **40**(1), 310–319 (2005)

113. Wang, C.C., Wu, C.F., Hwang, R.T., Kao, C.H.: Single-ended SRAM with high test coverage and short test time. IEEE J. Solid-State Circuit **35**(1), 114–118 (2000)

114. Wann, C., Wong, R., Frank, D., Mann, R., Ko, S.B., Croce, P., Lea, D., Hoyniak, D., Lee, Y.M., Toomey, J., Weybright, M., Sudijono, J.: SRAM cell design for stability methodology. In: IEEE VLSI-TSA International Symposium on VLSI Technology, 2005. (VLSI-TSA-Tech), pp. 21–22. 25–27 April 2005. http://ieeexplore.ieee.org/stamp/stamp.jsp? arnumber=01497065

115. Wittmann, R., Puchner, H., Ceric, H., Selberherr, S.: Impact of random bit values on nbti lifetime of an SRAM cell. In: 13th International Symposium on the Physical and Failure Analysis of Integrated Circuits, 2006, pp. 41–44 (2006). doi:10.1109/IPFA.2006.250993

116. Wong, V., Lock, C., Siek, K., Tan, P.: Electrical analysis to fault isolate defect in 6T memory cells. In: IEEE IPFA, pp. 101–104 (2002). http://ieeexplore.ieee.org/xpls/abs_all. jsp?arnumber=1025622

117. Yamaoka, M., Osada, K., Kawahara, T.: A cell-activation-time controlled SRAM for low-voltage operation in DVFS SoCs using dynamic stability analysis. In: 34th EuropeanSolid-State Circuits Conference, 2008. ESSCIRC 2008, pp. 286–289 (2008). doi:10.1109/ESSCIRC.2008.4681848

118. Yang, S., Powell, M., Falsafi, B., Roy, K., Vijaykumar, T.: An integrated circuit/architecture approach to reducing leakage in deep-submicron high-performance i-caches. In: The 7th International Symposium on High-Performance Computer Architecture, HPCA. 2001, pp. 147–157 (2001). doi:10.1109/HPCA.2001.903259

119. Yoshimoto, M., Anami, K., Shinohara, H., Yoshihara, T., Takagi, H., Nagao, S., Kayano, S., Nakano, T.: A divided word-line structure in the static ram and its application to a 64k full CMOS ram. IEEE J. Solid-State Circuit **18**(5), 479–485 (1983)

120. Zhai, B., Hanson, S., Blaauw, D., Sylvester, D.: A variation-tolerant sub-200 mv 6-T subthreshold SRAM. IEEE J. Solid-State Circuit **43**(10), 2338–2348 (2008)

121. Zhang, K., Hose, K., De, V., Senyk, B.: The scaling of data sensing schemes for high speed cache design in sub-0.18 m technologies. In: Symposium on VLSI Circuits, 2000, Digest of Technical Papers, Honolulu, pp. 226–227 (2000)

122. Zhang, K., Bhattacharya, U., Chen, Z., Hamzaoglu, F., Murray, D., Vallepalli, N., Wang, Y., Zheng, B., Bohr, M.: A 3-ghz 70 mb SRAM in 65 nm CMOS technology with integrated column-based dynamic power supply. In: IEEE International Solid-State Circuits Conference. Digest of Technical Papers. ISSCC, 2005, vol. 1, pp. 474–611 (2005). http://ieeexplore.ieee. org/xpls/abs_all.jsp?arnumber=1494075

123. Zhao, W., Cao, Y.: New generation of predictive technology model for sub-45 nm design exploration. In: ISQED '06: Proceedings of the 7th International Symposium on Quality Electronic Design, pp. 585–590. IEEE Computer Society, Washington (2006). doi:http://dx. doi.org/10.1109/ISQED.2006.91

124. Zhao, W., Cao, Y.: New generation of predictive technology model for sub-45 nm design exploration. In: 7th International Symposium on Quality Electronic Design, 2006, ISQED '06, San Jose, p. 6, p. 590 (2006)

Index

J. Singh et al., *Robust SRAM Designs and Analysis*, DOI 10.1007/978-1-4614-0818-5,
© Springer Science+Business Media New York 2013